John Bland-Sutton

Evolution and Disease

John Bland-Sutton

Evolution and Disease

ISBN/EAN: 9783744716611

Printed in Europe, USA, Canada, Australia, Japan

Cover: Foto ©berggeist007 / pixelio.de

More available books at **www.hansebooks.com**

EVOLUTION AND DISEASE.

BY

J. BLAND-SUTTON.

With 136 Illustrations

LONDON:
WALTER SCOTT,
24, WARWICK LANE, PATERNOSTER ROW.

1890.

CONTENTS.

	PAGE
INTRODUCTION	1

CHAPTER I.
THE ENLARGEMENT OF PARTS FROM INCREASED USE, OVERGROWTH AND IRRITATION ... 14

CHAPTER II.
DISUSE AND ITS EFFECTS ... 35

CHAPTER III.
VESTIGIAL PARTS ... 60

CHAPTER IV.
VESTIGIAL PARTS (*continued*) ... 79

CHAPTER V.
DICHOTOMY ... 102

CHAPTER VI.
ATAVISM OR REVERSION ... 134

CONTENTS.

CHAPTER VII.
ATAVISM (*continued*)—SUPERNUMERARY DIGITS, LIMBS, MAMMARY GLANDS 158

CHAPTER VIII.
PAGE
THE TRANSMISSION OF MALFORMATIONS AND ACQUIRED DEFECTS 176

CHAPTER IX.
ANATOMICAL PECULIARITIES OF THE TEETH IN RELATION TO INJURY AND DISEASE 199

CHAPTER X.
CAUSES OF DISEASE—INFLAMMATION AND FEVER ... 213

CHAPTER XI.
TUMOURS AND CANCERS 228

CHAPTER XII.
THE ZOOLOGICAL DISTRIBUTION OF DISEASE ... 250

INDICES 279

LIST OF ILLUSTRATIONS.

GOLDEN HEN-PHEASANT IN COCK'S PLUMAGE ... *Frontispiece*

FIG.		PAGE
1.	A DACE WITH SPOTS OF BLACK PIGMENT DUE TO THE IRRITATION OF A PARASITE	4
2.	HEAD OF FEMALE MOOSE WITH ANTLERS ...	5
3.	ANTLERS OF ROE-DEER WITH NODE	7
4.	HEAD OF COW WITH AN ABNORMAL HORN ...	8
5.	HEAD OF RHINOCEROS...	9
6.	THE LEG OF AN OYSTER-CATCHER WITH HORN	10
7.	HEAD OF THE HORNED PUFFIN...	11
8.	SECOND TOE ENLARGED FROM EXCESSIVE USE...	15
9.	FINGER OF HORSE AND MAN	18
10.	OVERGROWN HOOF OF GOAT	20
11.	HEAD OF COCK WITH SPUR	22
12.	SPINA BIFIDA OCCULTA	23
13.	HEAD OF POLISH FOWL	24
14.	HEAD OF DUCKLING WITH FOOT	25
15.	STOMACH OF DARTER	27
16.	A FISH EMBEDDED IN PEARL	30
17.	THE OWL-PARROT	36
18.	HEAD AND SUCKING DISC OF LEPIDOSTEUS ...	37
19.	ASCIDIAN AND ASCIDIAN TADPOLES	39
20.	THE HEAD OF VARANUS	42

LIST OF ILLUSTRATIONS.

FIG.		PAGE
21.	BRAIN OF HATTERIA	43
22.	A SECTION OF THE PINEAL EYE OF HATTERIA	44
23.	OVIDUCT IN A MALE SKATE	47
24.	THE EMBRYONIC ALIMENTARY CANAL	49
25.	DIAGRAM OF THE ALIMENTARY CANAL AND NERVOUS SYSTEM	50
26.	AN AFRICAN CHILD WITH A TUMOUR	53
27.	A FAUN	54
28.	AN ÆGIPAN AND FAUN	55
29.	OVERGROWN NAIL IN A SLOTH	57
30.	HEAD OF A PARROT WITH OVERGROWN BEAK	58
31.	A HORNED SHEEP WITH CERVICAL AURICLES	63
32.	THE CLOACA OF A HEN	64
33.	VERMIFORM APPENDIX OF A GIBBON	66
34.	FOLLICULAR CYST IN A PORCUPINE	72
35.	AN ODONTOME IN A HORSE	73
36.	AN ELASMOBRANCH FISH WITH ITS YOLK-SAC	76
37.	DIAGRAM OF THE ALIMENTARY CANAL AND YOLK-SAC	77
38.	AN EARLY HUMAN EMBRYO	81
39.	DIAGRAM INDICATING SITUATIONS OF BRANCHIAL SLITS	82
40.	GIRL WITH CERVICAL AURICLES	83
41.	A CHILD WITH CERVICAL AURICLES	83
42.	A COMMON GOAT WITH CERVICAL AURICLES	84
43.	AN EGYPTIAN GOAT WITH CERVICAL AURICLES	85
44.	MICROSCOPIC STRUCTURE OF A CERVICAL AURICLE	86
45.	SECTION OF CERVICAL AURICLE FROM A GOAT	87
46.	HEAD OF A SEAL	89
47.	HEAD OF AN ÆGIPAN WITH CERVICAL AURICLE	90
48.	FAUN AND GOAT FROM THE CAPITOL	91
49.	CONCRETION FROM THE GUTTURAL POUCHES OF A HORSE	95

LIST OF ILLUSTRATIONS.

FIG.		PAGE
50.	MALLEOLI OF MAN AND CHIMPANZEE	99
51.	SUPERNUMERARY RAYS IN STAR-FISH	103
52.	FEATHERS AND AFTERSHAFTS	105
53.	GEMINATED TEETH	106
54.	REDUPLICATED ANTLER OF MOOSE	107
55.	SUPERNUMERARY DIGITS IN A CHILD	108
56.	SUPERNUMERARY DIGITS IN A GIBBON	109
57.	A DOUBLE HAND	110
58.	MUD-FISH WITH DICHOTOMIZED LIMB	111
59.	SUPERNUMERARY HIND-LEG IN A TOAD	111
60.	SUPERNUMERARY FORE-LEG IN A FROG	112
61.	SUPERNUMERARY LEG IN A CHICK	113
62.	SUPERNUMERARY WING IN A DOVE	114
63.	SUPERNUMERARY FORE-LEG IN A SHEEP	115
64.	SECTIONS OF EMBRYO WORMS	117
65.	A TWO-HEADED COLT	119
66.	A DOUBLE SHARK	121
67.	A DOUBLE-HEADED SNAKE	122
68.	A MONSTROUS CALF	123
69.	A TWO-TAILEDED LIZARD	124
70.	A SIX-LEGGED FROG	126
71.	A SIX-LEGGED LAMB	127
72.	A FOUR-LEGGED CHICK	128
73.	A FIVE-LEGGED FROG	129
74.	A FOUR-LEGGED DRAKE	130
75.	A MITRAL VALVE CONTAINING MUSCLE	136
76.	A HAIRY MAN	138
77.	A ROCK LOBSTER (CEPHALON)	140
78.	PILIFEROUS CORNEA OF OX	142
79.	PATAGIUM IN THE LEG OF A GIRL	143
80.	WEBBED FINGERS IN A MONKEY	145
81.	HOUR-GLASS CONTRACTION OF THE STOMACH	147

LIST OF ILLUSTRATIONS.

FIG.		PAGE
82.	FEMALE ROE-DEER WITH ANTLERS	155
83.	HEAD OF A FEMALE MOOSE WITH ANTLERS	156
84.	MANUS OF HIPPARION	160
85.	DICHOTOMY OF A HORSE'S DIGIT	161
86.	DWARF LEMUR	164
87.	BRACHIAL MAMMA OF HAPALEMUR	165
88.	*LEMUR MACACO* AND YOUNG	166
89.	SUPERNUMERARY NIPPLES IN MAN	168
90.	SUPERNUMERARY NIPPLES IN A MONKEY	169
91.	INGUINAL RECESSES IN A LAMB	173
92.	MARSUPIUM AND NIPPLES OF A PHALANGER	174
93.	THE HUMAN PINNA	177
94.	A MALFORMED AURICLE	178
95.	SO-CALLED SUPERNUMERARY AURICLES	180
96.	DEVELOPMENT OF THE AURICLE	181
97.	DEFECTIVE PINNÆ	183
98.	SO-CALLED TAILLESS TROUT OF ISLAY	186
99.	TAIL OF A NORMAL TROUT FOR COMPARISON	186
100.	THE NOSE OF A HARE	189
101.	THE NOSE OF A DOG	189
102.	CLEFT LIP AND NOSE OF A DOG	190
103.	PART OF THE SKULL OF A DOG WITH A CLEFT PALATE	191
104.	HEAD OF A HUMAN EMBRYO OF THE FIFTH WEEK	192
105.	AN EXOSTOSED RAY FROM CHÆTODON	195
106.	HORNED MAN (SO-CALLED)	197
107.	INCISORS OF THE KANGAROO	200
108.	INCISORS OF KANGAROO SHOWING EFFECTS OF INJURY	200
109.	OVERGROWN TUSKS OF BOAR	203
110.	ERRATIC INCISORS OF BABIRUSSA	204
111.	BULLET IN AN ELEPHANT'S TUSK	207

LIST OF ILLUSTRATIONS. xiii

FIG.		PAGE
112.	THE ROSTRUM OF MESOPLODON	210
113.	SECTION OF A PYTHON'S TOOTH	211
114.	LEUCOCYTES ATTACKING BACILLI	218
115.	LEUCOCYTES INGESTING BACILLI	219
116.	CYSTIC KIDNEY OF A TERRIER	229
117.	THE CLOACA OF A HEN WITH CYST	231
118.	TRACHEAL POUCH OF THE EMU	232
119.	AN ACTINOMYCES TUFT	235
120.	SARCOMA IN THE NECK OF A FOWL	237
121.	MICROSCOPICAL CHARACTERS OF A SARCOMA	239
122.	MICROSCOPICAL CHARACTERS OF AN EPITHELIAL TUMOUR	243
123.	CANCER IN A PHALANGER	247
124.	ARTICULAR CARTILAGE OF A PIG WITH GUANIN GOUT	257
125.	GOUTY FOOT OF A PARROT	258
126.	AN ENDEMIC CRETIN	260
127.	A SPORADIC CRETIN	261
128.	A CALF-CRETIN	262
129.	PTAH	264
130.	MOUSE WITH CUTANEOUS HORN	267
131.	SECTION OF HORN	267
132.	HEAD OF A MOUSE WITH WART HORN	268
133.	HEAD OF A SHEEP WITH WART HORN	269
134.	HEAD AND LEG OF A THRUSH WITH WART HORN	270
135.	HEAD OF A LEPER	272

PREFACE.

———o———

My object in writing this book is simply to indicate that there is a natural history of disease, as well as of plants and animals. I have not attempted to deal with the subject at great length, far less exhaustively, but merely to illustrate general principles by a few carefully selected examples.

The subject is a novel one, and doubtless a more extended study will serve to show that many of my conclusions are fallacious. I trust that such errors may be speedily rectified by any person who has opportunities of testing my opinions practically. It must be borne in mind that it is a much easier task to write concerning the habits of animals than to describe their diseases; nevertheless, the facts at our disposal clearly indicate that disease is controlled by the same laws which regulate biological processes in general.

J. B. S.

EVOLUTION AND DISEASE.

INTRODUCTION.

MOST persons believe Pathology, as the Science of Disease is called, to be so outside the comprehension of ordinary individuals, and even in its general bearings so utterly devoid of interest to all but medical men, that much misconception prevails in the minds of even educated persons in regard to its fundamental principles. As a matter of fact Pathology is only a department of Biology, and it is very important to bear this in mind if we wish to study successfully the origin, cause, and spread of disease. Yet paradoxical as it seems, whilst so many regard Pathology as occupying an isolated position among sciences, medical writers always point out the difficulty they find in framing a definition of disease, and indeed the impossibility of stating where health ends and disease begins.

It is not my object in the present work to attempt the framing of a definition of disease, or even to offer a suggestion as to the borderland between it and health. This difficulty is frequently illustrated in a striking manner in a law court; it is not uncommon for a judge in the course of a criminal trial to ask a medical witness, when the plea of insanity is urged on a prisoner's behalf, either to define insanity, or to state his opinion where sanity ends and insanity begins. The judge knows full well the difficulty, indeed the impossibility of even a skilled witness making a satisfactory reply to such a question.

As with mental so with bodily conditions, it is impossible to state definitely the borderland between health and disease, either in relation with functional aberration or textural alteration. And in many instances we shall find conditions which we regard as abnormal in man, presenting themselves as normal states in other animals.

If it be difficult to define disease when our remarks are restricted to the human family, it becomes obviously more difficult when we attempt to investigate disease on a broad zoological basis. As the great barrier which exists between man and those members of his class most closely allied to him consists, not in structural characters, but in mental power, it necessarily follows that there should be a similarity in the structural alterations induced by diseased conditions in all kinds of animals, allowing, of course, for the differences in environment. This we now know to be the case, and it is clear that as there has been a gradual evolution of complex from simple organisms, it necessarily follows that the principles of evolution ought to apply to diseased conditions if they hold good for the normal, or healthy, states of organisms: in plain words there has been an evolution of disease *pari passu* with evolution of animal forms. For a long time it has been customary to talk of physiological types of diseased tissues, and my earlier efforts were directed in searching among animals for the purpose of detecting in them the occurrence of tissues, which in man are only found under abnormal conditions. The search was of great value to me, for the statement proved to be true in only a limited sense; at the same time the truth of an opinion held by nearly all thoughtful physicians, that disease may in many instances be regarded as exaggerated function, was forcibly illustrated, and I quickly saw that the manifestations of disease were regulated by the same laws which govern physiological processes in general, and that many conditions regarded as pathological in one animal are natural in another. It will be useful to illustrate this by some concrete examples. To take a simple case. The inside of our cheeks has a soft lining known as mucous membrane. In very rare instances children have been born with tufts of hair growing in this situation. Such a condition is truly abnormal. A physiological type for such a phenomenon is found in the mouths of rodent mammals; the inside of the cheeks of rabbits, hares, porcupines, and the like, present naturally patches of hairy skin. Pigment is widely diffused in animal bodies, both under natural and unnatural conditions, using the term unnatural as equivalent to disease; this explanation is necessary, for disease being controlled by natural conditions cannot logically be regarded as unnatural.

In the dace (fig. 1) we notice sundry collections of black pigment dotted among the scales. When examined critically the centre of each dot contains a white speck. These collections of pigment are due to the irritation caused by the presence of a parasite. In tigers, lions, monkeys, and sheep, similar pigmented spots are occasionally found in the lungs around parasites. In man, horses (especially grey horses), and dogs, tumours of an inky-black colour, called in consequence melanotic, are occasionally met with. All these formations of pigment are purely pathological. Under normal conditions, however, cuttle-fish (*Octopus*,

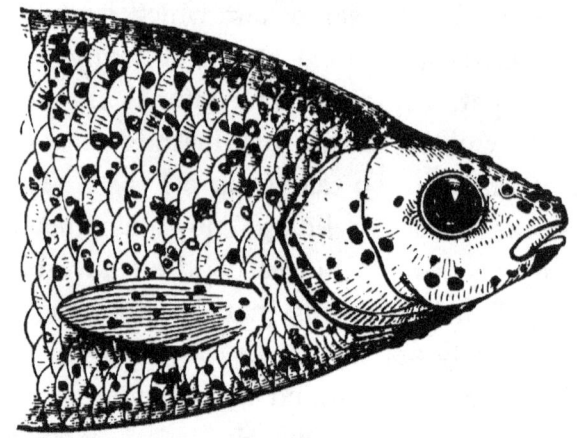

FIG. 1.—A Dace with spots of black pigment due to the irritation of a parasite (Mus. Royal College of Surgeons).

and *Sepia*), possess an ink-bag from which, when these animals are irritated, an ink-like pigment, sepia, can be ejected in such abundance as to colour the surrounding water to the extent of a cubic yard or more, and under cover of this dark cloud the cuttles escape from their enemies.

The close relation existing between physiological and pathological processes is shown in an interesting manner

INTRODUCTION.

by a study of the development and fall of the antlers of deer when compared with changes which occur in bone as a result of injury.

Bones are clothed externally by a membrane termed periosteum; this membrane serves as a matrix in which blood-vessels ramify before entering the compact tissue of the bone. It must be remembered that bone

FIG. 2.—The head of a female Moose (*Alces machlis*); the antlers are in "velvet."

is not only dependent on the periosteum for nutrition, but the deeper layers of this membrane have bone-forming properties; the increase in thickness of a long bone is due entirely to the periosteum. Should the periosteum be injured and inflammation established, a local increase in its bone-forming function is the result, producing a rounded or irregular swelling termed a node. In some cases the periosteum is so damaged that it becomes detached, and as a consequence the bone beneath dies. As soon as a piece of bone is dead those parts of the living bone adjacent become unusually active, leucocytes or white-blood cells begin to devour and finally succeed in detaching the dead portions when large, or digest them completely when small. Dead bone is known by the following features—it has no sensation, emits a sound when struck with a metallic instrument, and does not bleed when cut.

The antlers of deer when young and growing are covered with a soft vascular membrane, beset with delicate downy hair and glands, termed the "velvet," which bears the same relation to growing antlers that periosteum holds to bone (fig. 2). As long as the antlers retain this velvet in a living condition they increase in length and thickness; when the antlers are actively growing they feel warmer to the hand than the rest of the body, resembling in this respect an inflamed part. When in "velvet" a stag is particularly careful not to knock the antlers, for they are very sensitive, and when so unfortunate as to bruise them, a node or swelling forms upon them in every way resembling nodes on other bones when injured. I have seen nodes on antlers, caused by blows, as large as oranges. This is illustrated in fig. 3, which is a drawing of a pair of antlers of a roe-deer preserved in the museum of the Royal College of Surgeons. The left antler is shorter than the right one and has an ossified node upon it as large as a Tangerine orange. After the

antlers have attained full dimensions it is difficult for
the circulation to be maintained through so thin a mem-

FIG. 3.—A pair of antlers from a Roe-deer (*Capreolus capræa*) with an ossified node upon them (Mus. Royal College of Surgeons).

brane as the "velvet," and as a consequence it shrivels
and peels off; the bone beneath is deprived of blood

and dies. The branches suffer first and then the beam. At this stage the antlers become formidable weapons, and the stag, instead of taking every precaution not to knock or bruise them, now fears nothing, for they are like dead bone, devoid of sensation. In time the necrosis extends along the antler until it reaches the pedicle, that part which is covered by the natural hairy skin of the deer; in due course a line of demarcation is formed by leucocytes, and the antler falls

FIG. 4.—The head of a Cow with a large cutaneous horn.

by a process exactly analogous to that by which a piece of dead bone is separated.

We may turn to the consideration of processes in disease which are dominated by the physiological processes peculiar to a particular animal, and illustrate this by reference to cutaneous horns, especially that form which arises from the modification of warts. Not infrequently in mammals and birds the free portions of

INTRODUCTION. 9

warts become transformed into a tissue identical with horn. Such a specimen is represented in fig. 4. It shows a large horn projecting from the forehead of a cow, the horn is fifty centimetres in length. It was obtained by the celebrated John Hunter, and is preserved in the museum of the Royal College of Surgeons. A careful examination of the horn and the

FIG. 5.—The head of the Rhinoceros, showing the nasal horns, physiological type of wart-horn.

material which occupied the cavity in the horn, indicate that it originated in a wart. Such horns are common in man, and have been known to attain a large size. A physiological type of such horns is furnished by the nasal horn of the rhinoceros which in its structure, connections, and mode of origin resembles in its main particulars the pathological horn on the head of the cow;

indeed we are fully justified in stating that this nasal horn of the rhinoceros is a gigantic wart (fig. 5). Professor Flower recently exhibited at the Zoological Society the skin from the head of a rhinoceros shot in Central Africa with three nasal horns. The accessory one measured twelve centimetres in height and more than forty-two in circumference. It was situated in the same line posteriorly to the normal horns. It was structurally a wart. That cutaneous horns should arise in oxen and other hollow-horned ruminants (*Cavicornia*) need not surprise us when we reflect that [the corneous caps of their natural horns are modified portions of the integument.

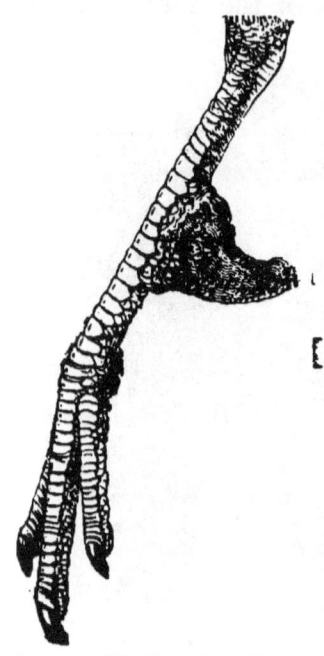

FIG. 6.—The leg of an Oyster Catcher (*Hæmatopus ostralegus*) with a large wart-horn.

Birds not infrequently exhibit wart-horns of this character, and an example growing from the leg of an oyster-catcher is shown in fig. 6. Such horns, whenever they have been observed in birds, follow the usual course of avian dermal organs in general, and are shed with each moult and reproduced with the new feathers. The horn on the leg of the oyster-catcher when compared with the size of the bird is very large. It is represented one-fourth its natural size.

The shedding of pathological cutaneous horns and their subsequent reproduction has more than one

INTRODUCTION.

physiological type. Among birds the horned puffin (*Fratercula corniculata*) will be selected.

Growing from the upper eyelid of this bird is a slender, pointed, black-coloured horn, which in the specimen from which the drawing (fig. 7) was made, measured eighteen millimetres in length: there is also a thin horny scale connected with the lower lid. In the adult bird it is stated that these horns are shed and reproduced annually.

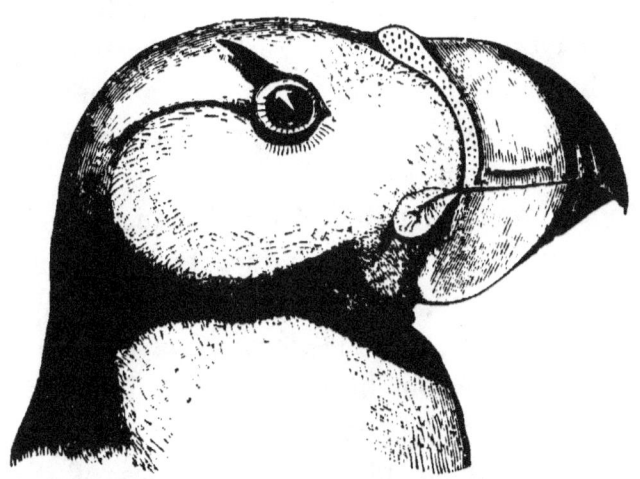

FIG. 7.—Head of the Horned Puffin (*Fratercula corniculata*) to show the horn growing from the upper eyelid.

It has already been mentioned that the corneous cap of the cavicorn ruminants is merely modified portions of the integument. In the Prongbuck (*Antilocarpa americana*) the hard cap of the horn is annually shed, an observation first made in 1865 in the Zoological Gardens, London. Subsequently, doubt was thrown on the matter, but it has been definitely settled by the observations of Mr. W. A. Forbes. Thus we are able to furnish types among normal cutaneous horns, not only in birds but among mammals, as parallels to the annual shedding of the pathological cutaneous horns of birds.

Not infrequently tumours are found in certain abdominal organs and in the subcutaneous tissues of man and other mammals, possessing skin and its appendages such as hair, wool, and glands. Such tumours contain in man, horses, and oxen, hair; in pigs, bristles; in sheep, wool; and in birds, feathers; thus harmonizing with the physiological characters special to the animal in which such tumours occur. Further, the hair in such tumour becomes grey as age advances, and like that on the exterior of the body is shed, so that such tumours may in the long run become literally bald.

Without attempting to multiply instances, such facts as these were sufficient to induce me to pursue the inquiry into Zoological Pathology, or General Pathology in the fullest sense, and the latest views and investigations in this wide, but little cultivated, field are summarized in the ensuing pages.

Wherever possible, physiological types of diseased (pathological) processes are described; the illustrations, whenever practicable, have been selected from animals other than man, for in him they have been too exclusively studied; indeed, by restricting our inquiries to man it is impossible to frame any generalizations concerning disease upon a sound basis. It has been stated that "a knowledge of human anatomy is sufficient for the mere art of the surgeon." This may be correct, but it is quite certain that if we restrict our observation of the processes of disease as they occur in man, our notion of them would be as crude as if we attempted to form conclusions as to his zoological position without reference to other species of animals.

In the ensuing chapters the following plan has been adopted: The effects of increased use and disuse of parts is considered in connection with the gradual change in the function of organs, and the part played by the transmission of the effects of increased use and disuse in producing vestigial structures in complex organisms. The tendency of vestiges to become diseased or to give rise to conditions disadvantageous to the individual is fully dealt with. The important and interesting subject of the transmission of acquired characters and malformations is briefly discussed, and a chapter is devoted to causes of disease arising without the organism and the relation they bear to the remarkable processes, inflammation and fever. Tumours are considered in connection with general morbid processes, and the scanty knowledge we possess of the zoological distribution of disease is summarized.

CHAPTER I.

THE ENLARGEMENT OF PARTS FROM INCREASED USE, OVERGROWTH, AND IRRITATION.

It is well established that the increased use of a part tends to enlarge and strengthen it, that disuse on the other hand often leads to its diminution and enfeeblement: structural modifications thus induced are inherited.

The truth of the first part of this statement may be demonstrated by a simple experiment: Let the arm of a healthy person be firmly strapped for several consecutive days upon a splint—in a few days the muscles will be softer than usual and actual measurements will show that the limb has diminished in size. Allow the arm to resume its function; the lost ground will be quickly recovered.

When a young and vigorous person has the misfortune to lose an arm the remaining limb, being used for all purposes, will rapidly increase in size and strength. The same facts may be observed in dogs and cats which have lost a limb or part of a limb. A woman, aged fifty, had her big toe, including the metatarsal bone, amputated; six months after she had regained the use of the foot, the second toe had enlarged, and stood out from its fellows in such a way as to resemble in size and general appearance the lost toe—indeed, when the foot was

ENLARGEMENT OF PARTS FROM USE.

exhibited to a class of students, this large second toe was mistaken for the hallux (fig. 8). This observation is of interest, the large size of the first toe and the great development of its muscles are owing to the greater use and importance of the hallux in mammals which maintain an erect, or semi-erect, position when walking along the ground as in man, or climbing trees as in monkeys and phalangers. Humphry, in reference to the large development of this toe, says, "Man literally stands in the animal world on his great toe."

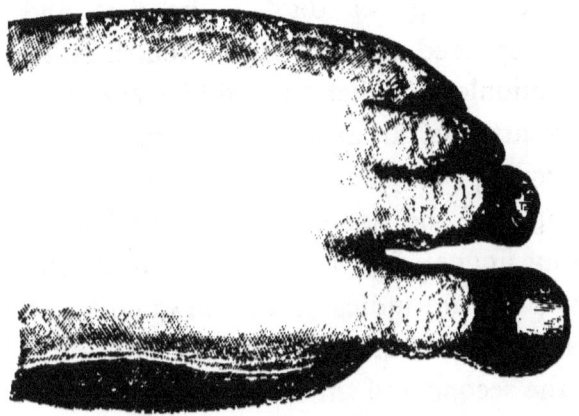

FIG. 8.—Enlargement of the second toe subsequent to amputation of the hallux.

The same remarks apply to the thumb: in man increased function develops its special muscles, thickens the bone, and toughens the nail. Even among quadrumana the pollex may be absent (Ateles); in such a high form as the chimpanzee the thumb is slender, short, and insignificant. In man we may attribute the disproportion of the hallux and pollex, in comparison with the neighbouring digits, to inheritance through a long line of ancestors of gradual increments of size induced by excessive use. Such gradual enlargement of a digit and its hereditary transmission may be demonstrated in *Equidæ*. The modern horse walks upon the greatly enlarged third digit of the hand and foot respectively, the hoof representing the nail. Hidden in the tissues on each side of this functional toe we find vestiges of the second and fourth; these are familiar to veterinarians as splint bones.

The researches of palæontologists have furnished an excellent array of evidence in support of the opinion that the horse has descended from ancestors which possessed five functional digits; the first and fifth gradually disappeared, the second and fourth still persist but are functionless, whilst the middle one has from increased use attained an extraordinary size.

In the drawing (fig. 9), a longitudinal section through this large digit of the horse is represented beside the corresponding finger of a man similarly bisected. The homologous parts are indicated by the same letters. In the horse's finger a bone is shown, in section, at the junction of the second and third phalanges; this is called the navicular bone or small sesamoid. Such bones are frequently found in the tendons of man, especially where they glide over bony prominences; usually they are small, rarely exceeding a split pea in size. In the horse such bones are large and important; not infrequently, when the foot is brought violently in contact with hard ground, the navicular bone in one or both feet is broken by the concussion; the result is permanent lameness, a fractured navicular bone rarely, if ever, unites by bone. In this respect it resembles the great sesamoid

ENLARGEMENT OF PARTS FROM USE.

bone in man, the patella or knee-cap; when the patella is broken by muscular violence it is rarely repaired by bone, but by yielding fibrous tissue.

In the navicular bone of the horse and the knee-cap of man, analogous conditions prevail, viz., bones which in many mammals are small and insignificant have become by excessive use enlarged and of such importance that, when damaged, permanent lameness in man, and uselessness in the horse, ensue.

When parts are enlarged in this way from increased use, they are said to be hypertrophied; in the case of the horse's foot the hypertrophy is said to be *functional*, in that of the big toe figured on page 15, *pathological*, as it arises in consequence of abnormal conditions.

The most striking examples of hypertrophy may be studied in the muscular system and in paired organs. For instance, should one kidney from any cause be slowly destroyed, the other will gradually enlarge and often double its size, thus compensating the animal and often preserving it from disaster. Many such cases have been reported in man and I have met with kidneys enlarged from this cause in horses, sheep, oxen, pheasants, and in a hen. Enlargement of a part from such causes is said to be *compensatory;* conspicuous examples of this form of hypertrophy occur in the animal kingdom. Variations in the size of a part according to the amount of work performed by it is illustrated by the gizzards of birds. In flesh- or fish-eating birds the muscular walls of the gizzard are relatively thin; in grain-eaters they are exceedingly thick. Hunter fed a sea-gull for a year on barley

18   EVOLUTION AND DISEASE.

and found the muscular coat increased in thickness. This simple experiment has been varied by other observers with similar results.

The large breast muscles (pectorals) of birds associated with and varying according to the expanse of wing, furnish a good example of the relation of increased size with augmented use ; these may be compared with the

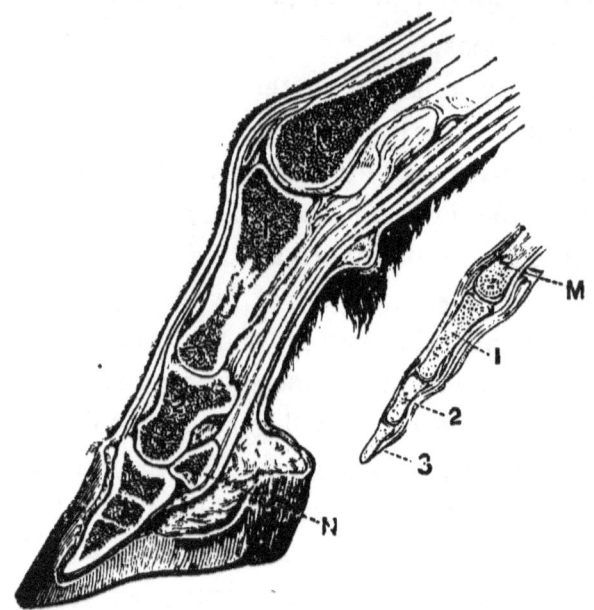

FIG. 9.—A medium longitudinal section through the third finger of a horse, and man. M, metacarpal bone ; 1, 2, 3, phalanges. The ossicle, N, situated at the junction of the second and third phalanges is the navicular.

powerful leg muscle of the frog (known as the gastrocnemius) so important in the act of swimming, and the large muscles of the human buttock, useful in helping man to maintain the erect position. These are striking instances of the inherited effects of increased use of a part.

ENLARGEMENT OF PARTS FROM USE.

Enlargement of parts may arise from increased blood supply due to irritation; thus a bone, the tibia, has been known to increase in length when long inflamed, to the extent of an inch and a half as compared with its fellow.

Skin supplies many curious and instructive instances under the names of corns and callosities. Those troublesome thickenings of the skin covering the toes, caused by ill-fitting boots, known as corns, or on the palms of the hands due to the use of tools in particular occupations, consist anatomically of a raised hard patch of thick epidermis; beneath it is a small sac containing fluid, termed a bursa. When a thickened patch of skin exists without a bursa it is usually called a callosity. Corns, as most are aware, occur most frequently on the toes, whilst callosities form on the sole of the foot and in the neighbourhood of the heel. Callosities are inherited, as is shown by the fact that the skin on the sole of the foot of a peasant's infant is thicker than that on the foot of the parson's offspring at the moment of birth. We may not unreasonably attribute the readiness with which a badly fitting boot will produce corns to a tendency we inherit from our parents and grandparents. In the same way the callosities on the breasts of camels, on the knuckles of the gorilla's fingers, and the ischial callosities of baboons, may be regarded as inherited local cutaneous thickenings, induced by the intermittent pressure to which the skin of these parts is subject; in the case of the camel when lying down, the gorilla when walking, and the baboon when sitting on its haunches.

Increase in the size of a part may arise from diminished use combined with irritation. For instance, nails, hoofs, and claws grow throughout life, and the wear and tear consequent on continued use is thus compensated. Should such parts be used less than usual, growth continues at the normal rate and the nails or hoofs become abnormally large and inconvenient. This form of enlargement is termed overgrowth, and is more liable to occur when diminished use is accompanied by irritation, as illustrated by the following specimen. It was a goat, confined for many weeks in a muddy paddock; on examining its feet I found them furnished with hoofs of great length, one measured thirty-six centimetres following the curve, and its fellow twenty-five centimetres (fig. 10). The hoofs of cows, horses, sheep, and deer become similarly overgrown when enclosed on marshy ground, dirty paddocks, or damp sheds. Similar conditions are not rare in our own species, for old, bedridden persons often have long toe-nails, some of them two or three inches long, thick and twisted like a ram's horn.

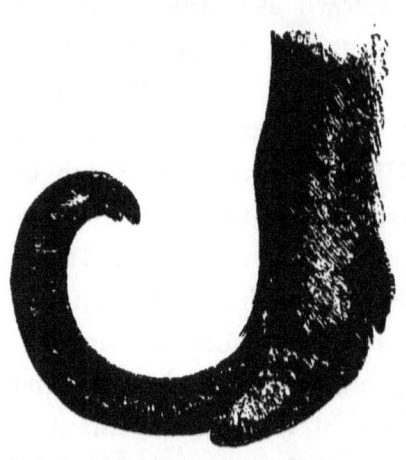

FIG. 10. — An overgrown hoof in a goat, which had lived many weeks in a muddy paddock. It measures thirty-six centimetres following the curve.

The relation between increase in the growth of nail or hoof in consequence of additional blood supply, is

ENLARGEMENT OF PARTS FROM USE.

shown in a striking manner in the foot of the horse. When inflamed the hoof will sometimes become much longer than usual, and the softer part, known as the frog, will enlarge and form a spongy mass, filling up the sole. This cannot be attributed to disuse as the remaining hoofs are also idle, for the horse is unable to work on account of lameness, and we can assure ourselves of the existence of inflammation in the affected foot by observing its increased heat.

In the next chapter I shall have occasion to demonstrate that overgrowth may occur from diminished use, alone; nevertheless, the most striking examples are associated with increased blood supply consequent on irritation, and as this assists in explaining some interesting normal conditions it will be advantageous to consider some additional instances. It has long been known that when the skin is irritated, especially in young people, by long continued discharges from wounds, or by the persistent application of poultices, the hairs of the part grow thick and long: when the irritation subsides the hairs gradually return to their normal condition. This overgrowth of hair may be attributed to a local increase in the blood supply, for it is a fundamental principle in Pathology that irritation produces redness of the skin; the heightened colour is a consequence of additional supply of blood to the part. Hunter demonstrated the relation between blood supply and overgrowth in an ingenious way: he transferred the spurs of cocks to the vascular tissue of the comb; here they took root and, in consequence of the extra supply of blood, and in part no doubt from disuse, grew inordinately. These specimens, in company with several others, are preserved in the museum of the Royal College of Surgeons (fig. 11).

The additional supply of blood to a part abnormally functional or irritated seems to be largely due to nervous influence, as the following experiments show. Bidder excised a piece of the sympathetic nerve in the neck of a young growing rabbit. This was followed by over-

FIG. 11.—The head of a cock with its spur transferred to the comb.

growth of the ear of the same side. The experiment has been repeated by Sterling on young and growing rabbits, and on dogs. A piece of the vagus and sympathetic—for in dogs both nerves are contained in the same sheath—was excised. In all cases the ear on this side became distinctly longer, broader, and somewhat thicker than its fellow. The hair was longer and

ENLARGEMENT OF PARTS FROM USE. 23

stronger on the side operated upon, and the ear remained distinctly warmer.

The relation of irritation upon nerves in connection with overgrowth of dermal structures may be illustrated by the curious defect known as spina bifida occulta. In this malformation the bony arches covering the spinal cord are defective, and the nerves issuing from the cord at this spot are involved in fibrous tissue or compressed by an accumulation of fat. It is no uncommon event to find the skin covering the defective parts of the spine presenting a tuft of hair often many centimetres in length, or the lower limbs may be covered with a crop of thick hair. The common form is shown in fig. 12.

These facts have been used in a subtle way by Virchow. The heads of Polish fowls are surmounted by a luxuriant tuft of feathers (fig. 13). Underlying this feathery crown in many Polish hens is a defect in the roof of the skull, resembling in many respects the condition known in man as meningocele. A study of the effects of spina bifida in man has led Virchow to regard the crown of feathers as the result of irritation, in the

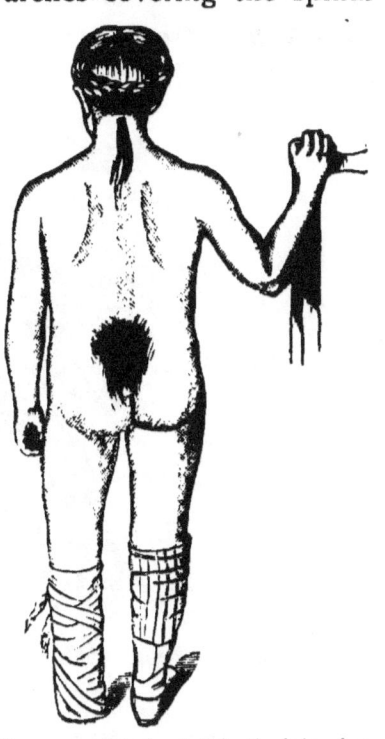

FIG. 12.—A hairy tuft in the loin, due to a defect in the arches of the spine and irritation of the cord or nerves. (After Fischer.)

same way that the hairy tuft may be accounted for in the back of those with spina bifida occulta. These fowls are extremely uncertain in their gait, given to performing circular movements, and walking sideways if excited, as though they possessed an unstable nervous system. Darwin was assured that we had here to deal with a character first acquired and transmitted by the hen.[1]

FIG. 13.—The head of a Polish fowl to show the feathery tuft.
(After Darwin.)

A somewhat similar condition is seen in ducks. Preserved in the museum of the Royal College of Surgeons is a duckling with a small tumour projecting from the top of its head; hanging from the side of the tumour is a miniature but well-developed foot (fig. 14). The swelling is connected with the duckling's brain by means of a small rounded hole in the summit of the cranium.

[1] "Animals and Plants under Domestication."

ENLARGEMENT OF PARTS FROM USE.

Isidore Geoffroy Saint-Hilaire [1] describes and figures the head of a duck with a tuft of feathers on the occiput and a foot. In describing the foot Saint-Hilaire states that the cranium underlying the tuft was defective, and especially notes that the cranial opening was similar to that found in "les poules à tête huppée."

FIG. 14.—Head of a duckling with a tumour and abnormal foot growing from the occiput.

During life this foot, like the normal pair, was of a beautiful orange-yellow colour. Some who saw the duck were suspicious that the foot had been engrafted on to the occiput accidentally: such an opinion had no foundation.

[1] "Des Anomalies de l'organisation chez l'homme et les Animaux," tome iii. p. 194.

Tiedemann [1] in 1831 described and figured the skull of a duck with a foot growing from its occiput. There is a fancy breed of ducks in which the distinguishing feature is the presence on the occiput of a rounded knob or swelling covered with feathers.

The acquisition and transmission of such characters shed light on some rather puzzling conditions. Abnormal growth of hair induced by contact with irritating substances may explain the presence of hair in such a curious situation as the stomach of a crayfish and the hairs which form the remarkable plug around the pyloric orifice of the darter's stomach (fig. 15). This bird feeds on fish, and as Garrod, in his excellent account of the anatomy of the darter's stomach, puts it, " This peculiar hairy mat acts as an excellent sieve to prevent the entrance of solid particles, fish-bones, &c., into the narrow intestines." [2] From what we know concerning the effects of irritation upon the skin it is quite conceivable that the contact of fish-bones and scales would act as irritants and induce a crop of hairs which, being advantageous to the bird, have been inherited. It in no way invalidates the argument by urging that skin, not mucous membrane, is furnished with hairs. Even the complex intestinal mucous membrane may, under exceptional circumstances, become converted into pilose skin. Such abnormal skin is more likely to possess hair if it be irritated. Abnormal growth of hair from irritation is paralleled by the elongation of the cutaneous papillæ under similar circumstances. This may be studied

[1] " Zeitschrift für Physiologie," Bd. iv. p. 121.
[2] " Proceedings of the Zoological Society," 1876, p. 335.

ENLARGEMENT OF PARTS FROM USE. 27

in lambs. In Britain sheep and lambs are often turned out to feed on clover grown in fields with the stubble from a previous crop of wheat remaining; the short, stiff, hard ends of the straw irritate the mouth and nose as well as the tender part of the feet above the coronet,

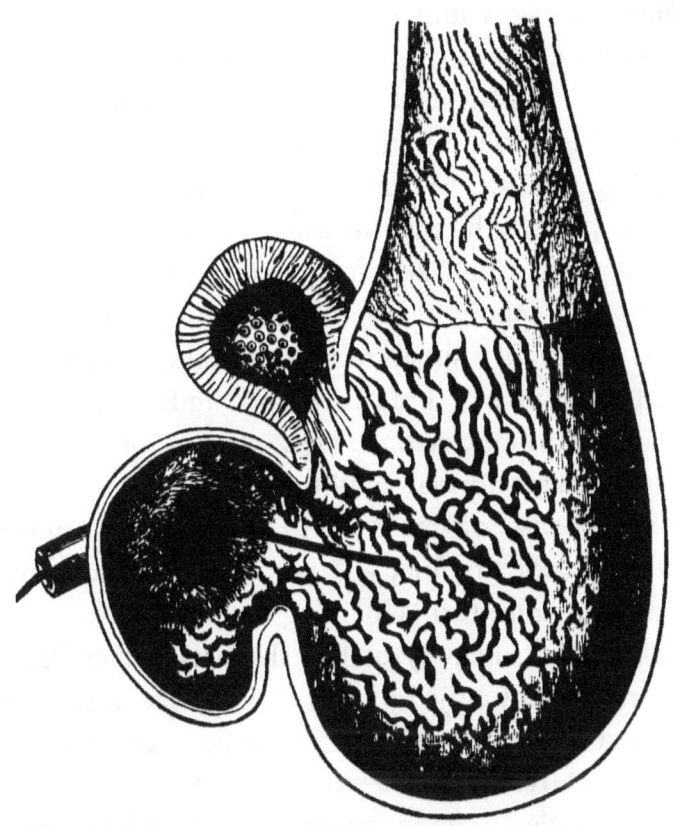

FIG. 15.—The stomach of the Darter (*Plotus anhinga*), showing the hairy pyloric plug (Mus. Royal College of Surgeons).

and produce a crop of warts. The relation of such irritation to warts is demonstrated by the fact that when the lambs are removed from the stubble the warts disappear. Similar warts grow on the hands of children when not kept clean; grubbing about in dirt and muddy, stagnant pools incidental to farmyards will produce warts on the noses of lambs.

The hairless pads on the feet of carnivorous mammals are made up of closely packed enlarged papillæ. When confined in cages and not kept scrupulously clean, the combined effects of dirt and limited use often induces a growth of warts. Sometimes, especially in the coatimundi, the whole of the pad will be covered with elongated papillæ, the appearance of such feet reminding us of the pad on the plantar aspect of the ostrich's toes.

The way in which skin responds to external stimuli explains the manifold modifications it presents in the various classes of animals, and it is highly probable that dermal structures of great utility to individual animals have arisen under circumstances such as produce them in man under abnormal conditions.

This response of the skin to irritation or abnormal stimulus is not confined to vertebrates; the lamellibranchs illustrate it in a striking way.

Lining the concavity of the shells is a membranous structure, which may be regarded as the integument, and is known as the pallium or mantle. The shell itself is the direct result of the excretory efforts of the lobes of the mantle, and is composed of animal matter hardened by deposits of carbonate of lime.

Occupying the space between the mantle of opposite sides, we find the animal proper, consisting of branchiæ, intestines, foot, nervous system, heart, reproductive organs, &c.

These animals obtain their food in a somewhat lazy

fashion. The margins of the gills are covered with cilia, which, by their constant movements, set up inhalent currents, which not only serve to oxidise the blood in the branchiæ, but convey concrete particles, many of which are seized upon by the mussel and utilised as food.

Some lamellibranchs have animals commensal upon them. Commensalism differs from parasitism in the important fact that an animal commensal on another lives upon the food of its host, whereas a parasite lives in the cavities or tissues of, and draws nourishment from, the blood of its host. It would seem that as long as the animals commensal on a lamellibranch keep within the space between the mantle they are safe enough, but occasionally they are rash enough to enter the space between the shell and the mantle. This trespass is resented by the lamellibranch, and the trespasser is punished by being entombed in shell-tissue, and in some cases by pearl.

A very beautiful example of this has been recorded by Dr. Günther.[1] The specimen is represented in the accompanying woodcut (fig. 16). It had been in Dr. Günther's possession for many years. It is an old shell of *Margarita margaritifera*, in which there is embedded, behind the impression of the attractor muscle, a perfect individual of a fish belonging to the genus *Fierasfer*. The fish is covered by a thin layer of pearl-substance, through which not only the general outlines of the body, but even the eye and mouth, can be seen.

In this case the fish, instead of keeping between the

[1] " Proceedings of the Zoological Society," 1886.

two halves of the mantle, penetrated between the mantle and the shell. The irritation thus caused induced the mollusc to cover the intruder with pearl. The secretion must have taken place in a very short time, at any rate before the fish could have been destroyed by decomposition.

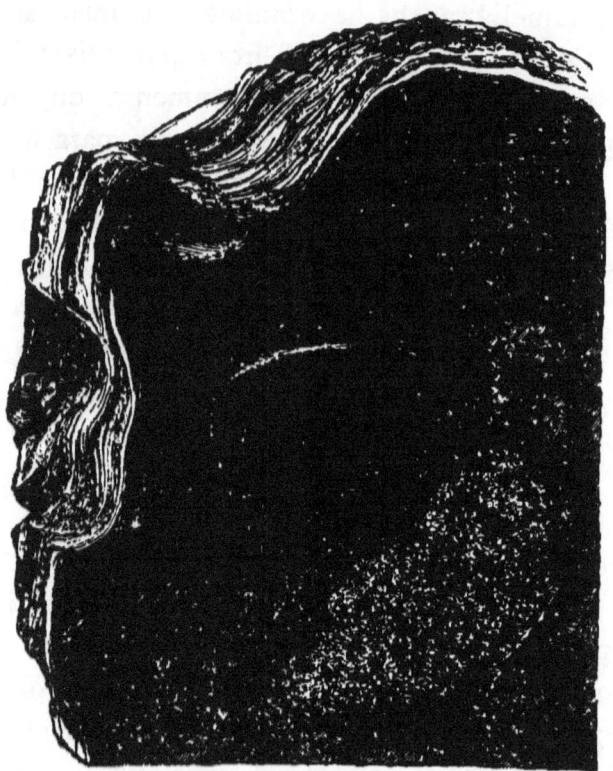

FIG. 16.—A fish embedded in pearl (after Günther).

Dr. H. Woodward has described some interesting specimens of the same nature. He exhibited before the Zoological Society (1886) a *Pinnotheres* which had been entombed in a cyst of pearl by a pearl-mussel. In connection with the specimen Dr. Woodward made the following remarks: "It seems extraordinary and beyond belief that the *Meleagrina* should, of all the Conchifera, be the one to resent the commensalism of the Pea-crab, which has been known since the days of Cicero, Pliny, Oppian, and Aristotle to inhabit the shell of the Pinna and the Oyster, and has been recorded from *Astarte, Pectunculus*, and at least some half-dozen other bivalves, with whom it appears to live on the most friendly terms. It is the females, however, which constantly reside within the shells of the Conchifera, whilst the males are said to avail themselves of favourable opportunities to visit the females in their retirement."

Whether or not the unlucky male in this case intruded himself upon *Meleagrina* at an unfavourable period, and, finding no female *Pinnotheres*, penetrated so far beneath the mantle of the pearl-mussel as to be unable to retreat, one thing is quite clear, namely, that *Meleagrina* entombed the intruder in a cyst of pearl, from which the clever pearl-button maker alone liberated him.

Increased thickness of the shell of an oyster from irritation is comparable to the formation of thick skin under similar conditions. The thin shells of "native oysters" living in quiet shallow water stand in striking contrast to the huge, rough, laminated shell of the "North Sea oyster" which has to contend with the pressure of a large volume of water and much buffeting from a tumultuous ocean. So that a native oyster, when compared with one from the North Sea, resembles the hand of a courtier when contrasted with that of a peasant.

That increased size is not due always to increased function but may arise from irritation is admirably illustrated in *Solanum jasminoides* described by Darwin. " The flexible petiole of a half or a quarter grown leaf which has clasped an object for three or four days increases much in thickness, and after several weeks becomes so wonderfully hard and rigid that it can hardly be removed from its support. On comparing a thin transverse slice of such a petiole with one from an older leaf growing close beneath, which has not clasped anything, its diameter was found to be fully doubled and its structure greatly changed " ("Climbing Plants").

In this example the extra thickness could not be due to increased function, but to irritation ; the petiole had less work to perform as the leaf was largely supported by the object which its petiole had clasped.

The effects of increased use may be observed in the organs of special sense. When an individual loses an eye in early life the remaining healthy eye acquires a greater range of movement and quickness which compensates in no small degree for the loss of its companion. In persons blind from early life the power of hearing becomes wonderfully quickened, and their tactile sensibility is so heightened that they make themselves acquainted with external surroundings in a marvellous manner. Similar instances are furnished by the mole ; its sense of hearing is proverbial. Says Caliban to Stephano and Trinculo : " Pray you tread softly, that the blind mole may not hear a footfall." The blind fish of the mammoth cave are said to be abnormally sensitive to sounds as well as to undulations produced

ENLARGEMENT OF PARTS FROM USE. 33

by various causes in the water. Wyman has attempted to show that the semicircular canals are unusually large in the blind fish *Amblyopsis*.

The study of the examples of enlarged parts arising from increased use, additional blood supply, and irritation, teaches clearly enough that the same laws which regulate these processes under normal conditions are equally active under abnormal conditions, and indicates that the thick fur of mammals living in cold climate, or the local growth of hair on the skin of man when stimulated by irritants or unusual states of the nerves, are responses to such stimuli as call up the growth of hair in the stomach of the darter or crayfish; the scales of serpents, feathers of birds, quills of porcupines, and bristles of hogs, are like hair, epidermis and horn, modifications of the surface epithelium, probably induced by variations in the nature of the stimuli, or irritants, and in differences of surrounding conditions, such modifications being transmitted to the offspring.

The inheritance of the effects of increased use of parts not only manifests itself in enlarged muscles, thick bones, and stout ligaments, but explains the large udders and bountiful supply of milk we obtain from domesticated cows, and, as Wallace rightly remarks, "almost perpetual egg-laying in poultry."

In a similar manner increased use of the special senses with the transmission of the extra acuteness gradually acquired by individuals, explains the wonderful power of scent in dogs, of sight in hawks, and of cunning in foxes.

In a similar manner, also, the use of the fingers in

particular trades, as those of watchmakers, jewellers, ivory carvers, &c., or the precision of artists and sculptors, and, in a more marked sense, the keen ear of musicians, and the wonderful faculty displayed by mathematicians in their marvellous dealings with numbers and calculations, have all been slowly attained by the persistent transmission of the effects of increased use.

CHAPTER II.

DISUSE AND ITS EFFECTS.

DISUSE of a part usually leads to its enfeeblement and diminution, a result conveniently expressed by the term atrophy; in many instances the effects of disuse are transmitted. Atrophy may be induced in a variety of ways, but in nearly all cases it is attributable to diminished use and its inevitable consequence, lessened blood supply. Disuse of a part may be caused by changed habits of life, or by the increasing importance of some other organ. Certain parts are only useful for a brief period in an animal's life; some appear to have no function and are present in conformity with the law of heredity, whilst atrophy from disuse may be the consequence of injury; and, lastly, an interesting variety of atrophy is due to continuous pressure. It will therefore be instructive as well as conducive to clearness to describe some typical cases of the various forms of atrophy.

Atrophy from changed habits.—Among the many anomalies of animal life in New Zealand must be included the remarkable owl-parrot or kakapoe (*Stringops habroptilus*). This bird is nocturnal in its habits, feeds on fern-shoot, roots, berries, and, it is said, occasionally lizards. It climbs but does not fly, though possessing what looks like, in so far as shape and size are concerned, an admirable pair of wings. A dissection of the pectoral muscles is suggestive, for they are thin, flat, and contain but little contractile tissue. The prominent keel so conspicuous on the sternum of flying birds is, in *Stringops*, a mere ridge. The intrinsic muscles of the wings are pale, thin, and composed largely of fibrous tissue (fig. 17).

It has been inferred that these birds have not long been inhabitants of New Zealand only, but were de-

FIG. 17.—The Owl-parrot, or Kakapoe (*Stringops habroptilus*).

veloped in other countries where their wings were of use to them. The disuse of the wings is due to alteration in environment.

The atrophied wing muscles in the owl-parrot recall the observations of Rengger who attributes the thin legs and thick arms of the Payaguas Indians to successive generations having passed nearly the whole of their lives in canoes with their lower extremities motionless.

Atrophy of parts useful for a brief period.—Very many organs are useful for a brief period, and later

DISUSE AND ITS EFFECTS. 37

atrophy or even disappear. The curious suctorial disc of the recently hatched embryo of the fish, lepidosteus, is a case in point. In the adult fish the upper jaw ends in a fleshy globular projection ; this, in the embryo, is a large disc, as in fig. 18. Agassiz, to whom we are indebted for much of our knowledge of this structure, has ascertained that the disc is formed two or three days before hatching, and the young fish uses it as a sucker,

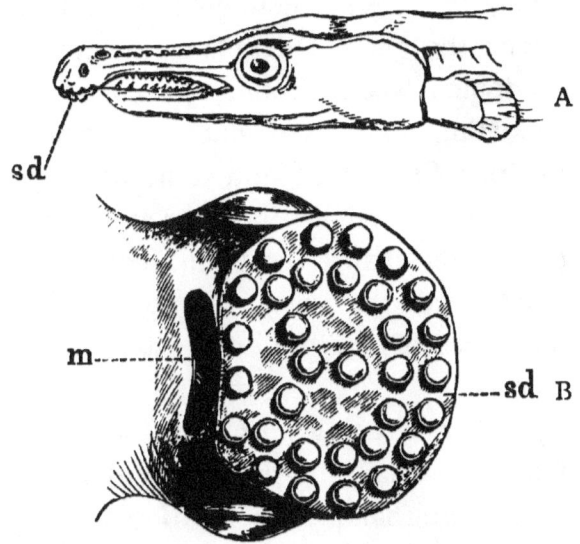

FIG. 18.—A, head of a young lepidosteus ; B, the suctorial disc seen from below ; sd, suctorial disc ; m, mouth. (From Balfour.)

by means of which it can attach itself to the sides of the vessel in which it is confined, or to other objects. The young lepidosteus can fix itself so firmly that considerable commotion in the water is required in order to make the fish lose hold : it can even remain suspended after the water has been lowered beyond the level to which it is attached.

This form of atrophy may be studied in ascidians. These are marine animals which encrust stones, rocks, and weeds on the sea-bottom. Sometimes they are solitary, but often occur combined in masses. In the adult condition they have an appearance recalling that of a tough leathern bottle with two openings; water enters at one, leaves by the other. The young of some ascidians have a totally different form, resembling a tadpole not only in external configuration but in internal organization; the general details of the anatomy of an ascidian tadpole is shown in fig. 19.

After existing in a free state for some time the young ascidian fixes itself to a stone by its head; the tail, with the notochord and nervous axis, atrophies, the body changes its shape, the brain remains small and undeveloped, and the eyes disappear. Finally the animal increases in size, its outer case becomes tough and leather-like.

Among other examples of this form of atrophy, mention may be made of the tail and gills of frog-tadpoles, the external gills of sharks, and the Alpine salamander, the yolk sac of vertebrata. Remarkable instances of the atrophy and disappearance of larval organs may be studied among invertebrates, especially in the echinoderms and star-fish. Many marvel at such things occurring in other animals, and overlook the fact that similar conditions may be studied in our own bodies, for the fall of the milk teeth is induced by the same process which brings about the disappearance of the tadpole's gills and tail. Puppies are born blind: this blindness is due to the existence of a vascular

DISUSE AND ITS EFFECTS. 39

membrane occupying the space known as the pupil of the eye. This membrane rapidly atrophies and the pups begin to see. In the human offspring a precisely similar membrane is present during embryonic life, but atrophies a few weeks before the termination of intra-uterine life. The mode by which fœtal or larval organs of this character are slowly removed will be considered in detail in a later chapter.

Suppression of Parts.—It has long been known that in the embryo of vertebrates

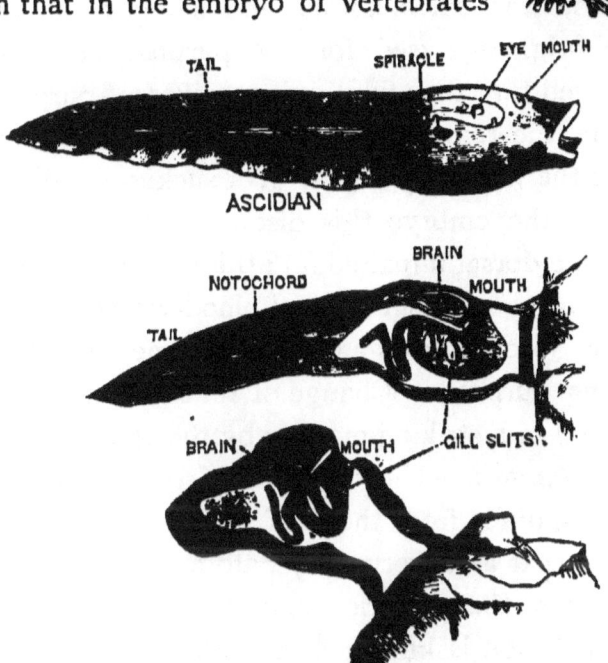

Fig. 19.—Ascidian Tadpoles (after Lankester). A, An adult ascidian.

many structures and organs are formed and then disappear without attaining, as far as our knowledge allows us to judge, a functional condition.

The only interpretation which can be put on this fact is, that these organs or parts have been functional in the ancestors of such animals, but in consequence of the increased use of other parts and change in surrounding conditions, the organs in question are not so serviceable to the animal and have, in consequence of diminished use, slowly but gradually atrophied. This effect may be conveniently referred to as suppression. It is also a point of some importance to remember that in consequence of changed conditions in the surroundings and habits of an animal, organs originally used for one purpose may become so changed that they fulfil quite a different purpose: for example, the remora, or sucking-fish, is able to attach itself to the shark by means of a sucker-like disc on its head; in the embryo this disc arises from the anterior part of the dorsal unpaired fin; this is indicated throughout life by the arrangement of blood-vessels and nerves. Thus a locomotor-organ has become modified for attaching purposes. Change of function and atrophy is illustrated in a striking manner by the allantois.

The Allantois.—The embryos of reptiles, birds, and mammals, differ from those of fish and amphibians by the fact that at a very early date a vascular dilatation arises from the posterior end of the developing gut. This dilatation is known as the allantois, and in birds and reptiles spreads itself beneath the shell-membrane. The blood circulating in the membrane is in this way brought into favourable relations with the atmosphere; the air diffusing through the shell, oxidises the blood in the allantoic capillaries. The allantois is the respiratory organ of the sauropsidian embryo. In the embryo of placental mammals the work of respiration is considerably modified, being performed by means of the placenta, an organ formed of structures derived in part from the fœtus and in part from the mother. Under these conditions the function of the allantois is limited to the conveyance of blood-vessels from the embryo in order to bring them into intimate relation with the maternal tissues. This work accomplished, the allantois shrivels, with the exception of the part in relation with the cloaca; this becomes permanently useful in mammals as the urinary bladder; a portion, however, remains as a withered cord passing from the summit of the bladder to the navel and is known as the urachus.

Thus the allantois exhibits what at first sight appears to be a change from a respiratory organ to a receptacle of urine, but closer inquiry shows the matter to be somewhat different. Amphibians possess a urinary bladder, but not an allantois: a critical inquiry into the matter induces me to accept Balfour's view and to look upon the allantois as an enormously enlarged urinary bladder which assumed in the embryo respiratory functions. This change is coincident with, if not responsible for, some extraordinary alterations. Fish and amphibia (*Ichthyopsida*) differ from reptiles and birds (*Sauropsida*) and mammals in that they possess during some period of their lives, gills, and in the non-possession of a functional allantois. No vertebrate is known which possesses gills and a functional allantois. Before the advent of the allantois, embryonic respiration is carried on in a variety of ways, sometimes by external gills, as in sharks and the Alpine salamander, or by means of the yolk sac forming adhesions to the oviducal wall, as in *Mustelus lævis*, or even by means of the tail (*Cæcilia compressicauda*).

It is not unjustifiable to hold the allantois responsible for the abolition of gills in sauropsida and mammalia. This is much more probable than to attribute the change to the evolution of lungs from a swim-bladder, for functional gills and lungs co-exist in such forms as the mud-fish (*Lepidosiren*) and ceratodus.

The history of the pineal eye is an instructive instance. Connected with the vertebrate mid-brain is a structure known as the pineal body, which has long puzzled anatomists. Many investigators have regarded it as vestigial; that is, it was of some functional value in the ancestors of existing vertebrata. The truth of this opinion has been demonstrated by the admirable researches of De Graaf and Baldwin Spencer.

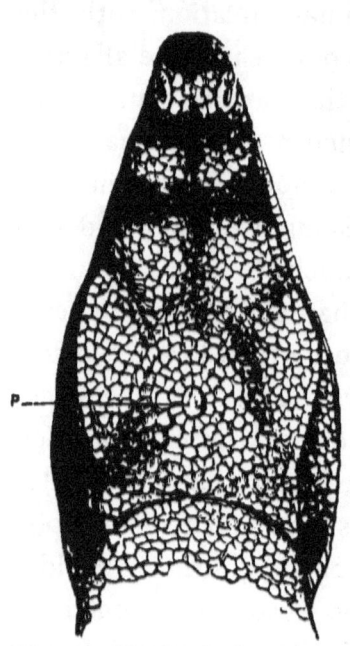

FIG. 20.—The head of a Lizard (*Varanus*), showing the bright scale indicating the pineal eye, P.

On the dorsal aspect of the skull in lizards a small opening exists, known as the parietal foramen. In some lizards — as, *e.g.*, *Varanus*—the situation of this foramen is indicated by a bright scale (fig. 20). On making a longitudinal section of the head, so as

DISUSE AND ITS EFFECTS. 43

to include this parietal foramen, we find it occupied by an organ representing an eye in miniature, connected with the pineal body by nerve-fibres. The relation of the parts is indicated in fig. 21, representing a lateral view of the brain and adjoining parts of the skull in the New Zealand lizard, Hatteria (*Sphenodon*).

When the eye is examined microscopically, it presents the structural details found in functional eyes, such as cornea, lens, retina, pigment, &c. (fig. 22).

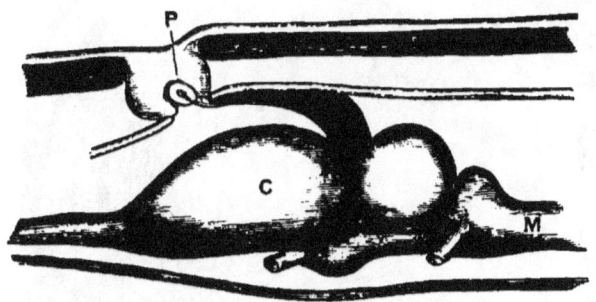

FIG. 21.—Lateral view of the brain of Sphenodon, showing the relation of the pineal eye, P; the cerebrum, C; and the medulla, M. (After Baldwin Spencer.)

Spencer is of opinion that this suppressed eye represents the unpaired eye of larval Tunicata. There is good evidence that it was highly developed in extinct amphibia (*Labyrinthodonta*), and was probably a sense-organ in animals of pre-tertiary periods.

It is not unreasonable to suppose that the gradual development and greater utility of the lateral eyes have led to the suppression of the median eye.

Although the pineal body (regarded by Descartes as the seat of the soul) in man is clearly vestigial, it is

44   EVOLUTION AND DISEASE.

by no means harmless, for occasionally it enlarges and becomes occupied by tumours, sometimes of large size and complex constitution, which cause death from mechanical interference with the brain.

Darwin has pointed out that rudimentary (vestigial) parts are apt to be highly variable. This variability he thought was largely due to uselessness, and there-

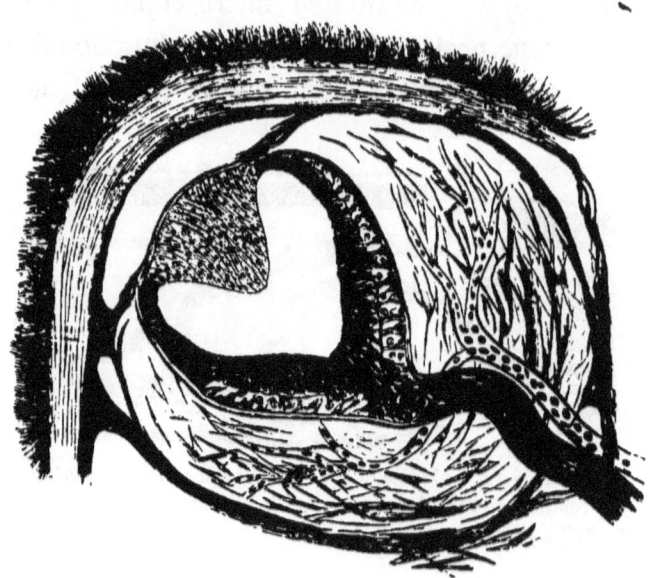

Fig. 22.—A magnified view of a section through the pineal eye of Hatteria. (After Baldwin Spencer.)

fore to natural selection having no power to check deviations in the structure of such parts.

It will be interesting to study this view of the matter in connection with oviducts, which are of frequent occurrence as anomalies in the males of unisexual animals, especially as two reasons can be advanced for their vagarious conduct. In the first place, I shall endeavour to show that they have undergone a remarkable change of function, inasmuch as they are transformed from urinary to reproductive functions in the female, whilst in the male they are functionless.

The Oviducts.—The history of the excretory organs indicates that the kidneys were larger in the ancestors of vertebrata than in existing forms, and the urinary excretion was conveyed by a series of ducts to the exterior of the body, whereas the kidney possesses now only one duct, the ureter. Of the once extensive renal system, the higher vertebrata possess relics in the form of the Wolffian bodies and their ducts, structures which are relatively very large in the embryo, but towards the mid-period of intra-uterine life dwindle, and are for the most part present in adults as vestiges. Although the glandular portions of the primitive renal system atrophy, and are permanently replaced by the kidney, the ducts belonging to them undergo a great change, and become utilized for reproductive purposes.

In some fish the ova are shed into the abdominal cavity from the ovary, and then escape to the exterior by small openings near the anus, known as genital pores. In many fish the ducts which belonged to the anterior section of the primitive kidney become modified into egg-conduits or oviducts.

When dealing with the Laws of Variation, Darwin states that "a part developed in any species to an extraordinary degree or manner, in comparison with the same part in allied species, tends to be highly variable." This law may be extended beyond the scope of species, and applied to the reproductive ducts, for we must regard them as being extraordinarily modified as well as highly variable. This may be illustrated by examples from widely separated animal forms. It will be advantageous to commence with the oviducts. In male toads and frogs two slender streaks may generally be detected passing from the so-called vesiculæ seminales forwards to the roots of the lungs. These thin streaks are the oviducts; in the male they are functionless, and normally remain diminutive. It is by no means uncommon to find, especially in male toads, the duct on one or both sides, as well developed as in the female.

Professor Howes has described some well-marked instances of the persistence of portions of the oviducts in male specimens of the green lizard, and in one the entire oviduct persisted as in the female.

One of the most noteworthy examples of a persistent oviduct in a male is that recorded by Mr. J. D. Matthews,[1] which he met with in a skate dissected in the Natural History Department of the Edinburgh University. In this fish a well-developed oviduct was found on the left side in association with male organs (fig. 23). An examination of the drawing shows clearly enough that this oviduct was not a mere rudiment, but was of the same proportion as would be found in a female skate of corresponding size. The claspers were present, and about six inches long.

This tendency of the oviducts to persist in the male is not limited to fish and amphibians, but is manifested

[1] *Journal of Anatomy and Physiology*, vol. xix. p. 144.

by sheep, deer, oxen, monkeys, goats, and occasionally in man.

In mammals the oviducts become transformed into a complex uterus. Other ducts belonging to the Wolffian bodies are modified in a similar way, serving to convey the products of the male generative gland to the exterior.

It is somewhat remarkable that each vertebrate embryo possesses male and female reproductive ducts; the adult male of most vertebrates possesses vestiges of the female ducts, whilst the adult female possesses, much more constantly, easily detected remnants of the male sperm-ducts. Remembering that the primitive renal organs are common to both sexes, and as the disused ureters have been utilized in the male and female for reproductive purposes, it renders their temporary co-existence in the embryo, and the persistence of one or other in a vestigial form according to the sex, comprehensible without invoking aid from the much-disputed question of the existence of a condition of primitive hermaphrodism.

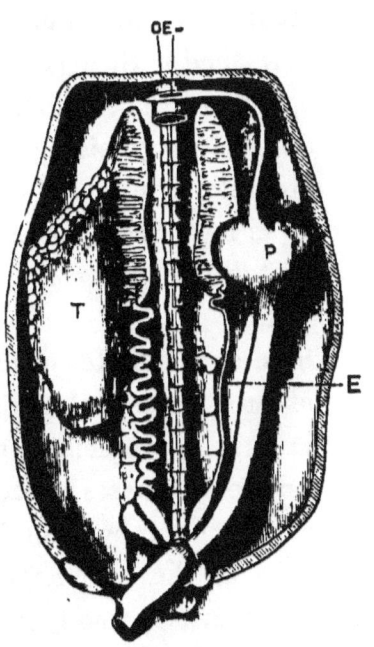

FIG. 23.—The male reproductive organs of a Skate, with a fully-developed oviduct on the left side. Œ, œsophagus; T, testis; E, epididymis (after J. D. Matthews); P, shell-gland.

The history of the reproductive ducts in the female

of vertebrata is certainly very remarkable, but a much more extraordinary instance of change of function is manifested in the nervous system now to be considered.

The Central Nervous System.—Vertebrate animals are distinguished from invertebrate not only in the possession of a vertebral column, but also by the fact that they are furnished with a central nervous system known as the spinal cord and brain. Of late years it has been the opinion of many biologists that the separation of the invertebrata from vertebrata is unjustifiable, and many attempts have been made to bridge the gulf supposed to exist between these great divisions of the animal kingdom.

Those invertebrates which approach nearest the vertebrates are the cephalopods (cuttles, octopods, &c.), and in these forms the central nervous system is represented by ganglionic masses collected around the œsophagus or gullet and united by commisural fibres : this arrangement is known as the œsophageal collar, so that in order to bring this into harmony with the anatomical disposition of the vertebrate gullet, some eminent biologists have maintained that the vertebrate mouth is secondary, and that the primitive gullet traversed the central nervous system by way of the third cerebral ventricle and infundibulum, a diverticulum from the original vesicles out of which the brain is ultimately developed. This hypothesis has not found much favour, but recently some observations and speculations have been announced which throw much new and important light on the matter.

The central nervous system is traversed by a canal

DISUSE AND ITS EFFECTS. 49

which is of large relative size in the embryo and lined by epithelium. The significance of this canal has long puzzled anatomists. A new interest attached to it when Kowalevski discovered in the embryo of ascidians and amphioxus, that this central canal is directly continuous with the intestine. This temporary connection

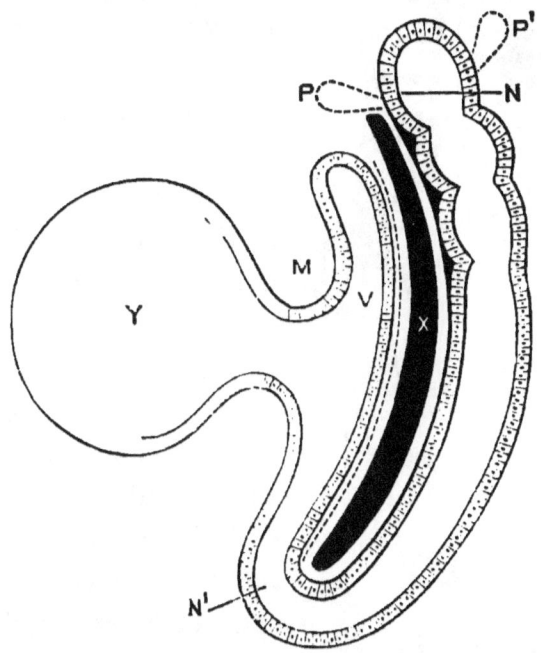

FIG. 24.—The U-shaped tube from which the alimentary canal and central nervous system of vertebrates arise. N, nervous tube; V, intestine; N^1, neurenteric passage; X, notochord; Y, yolk sac; P^1, pineal; and P. Pituitary diverticulum.

has been observed in all the great groups of vertebrata. even in the human embryo.

In 1887 I was able to furnish evidence that this central canal of the cord, and a portion of it prolonged into the brain, may be regarded as originally a segment of intestine which has become disused for alimentary purposes

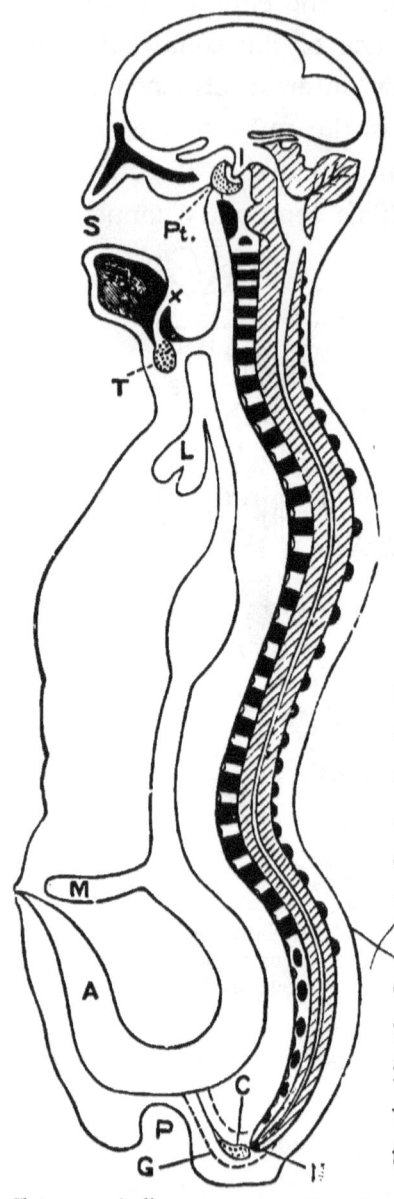

and undergone gradual transformation into a spinal cord. These facts were based on the embryological history of the parts and from a study of the malformation exhibited, not rarely, by this central canal and the permanent alimentary canal.

The mode of development of the parts may be briefly described. The alimentary canal and spinal cord of all vertebrates arise on a common plan. At a certain stage it consists of a U-shaped tube, as in fig. 24, each limb ends in a *cul-de-sac*, and the ventral limb is connected by a hollow duct with the yolk sac.

By a series of secondary changes the anterior end of the ventral limb of this tube is made to communicate with the exterior by way of the mouth and pharynx, and

FIG. 25.—A diagram representing the parts in the adult derived from the U-shaped tube of the embryo. S, stomodœum; I, infundibulum; Pt, pituitary body; T, thyroid; L, lung; M, vitello-intestinal duct; A, allantois; C, coccygeal body; G, post-anal gut; N, neurenteric passage.

by a similar method a nether opening, the anus, is established. The intermediate section becomes the permanent alimentary canal. The walls of the dorsal section of the tube thicken, the cavity becomes restricted, and the bend connecting it with the ventral tube atrophies, thus disconnecting the two limbs; the dorsal portion finally developes into the brain and spinal cord. The various connecting parts are represented in the adult by the following structures, diagrammatically indicated in fig. 25. The diverticulum, P, is the infundibulum; this was closely associated with the primitive gullet. The connecting limb, N, completely disappears, but the section of the gut into which it opens is represented by a small pedunculated body at the extreme end of the vertebral column, and known as the coccygeal gland. The dorsal and ventral limbs of these tubes in the adult are connected in a subtile manner by means of nerves. The walls of the dorsal tube contain collections of nerve-cells, from which nerves issue, portions of which are distributed to the body walls, others of peculiar character ramify in the walls of the intestines and are intimately associated with its nerve plexuses.

This view as to the intestinal origin of the central canal of the nervous system receives admirable support from the investigations of Dr. Gaskell,[1] who, from a

[1] My conclusions were framed in July, 1887, and briefly stated to the Pathological Society in October of that year. My manuscript was sent to "Brain" in August, but owing to the unfortunate illness of the Editor it was not published till January, 1888. Gaskell states that he framed his conclusions in the summer of 1887, but he did not mention them in public till June, 1888, and did not publish them till April, 1889.

different mode of working to that adopted by me, has come to a similar conclusion. Gaskell's conclusions are based mainly on the fact that the central nervous system is composed of two parts, a nervous and a non-nervous element, especially in the cranial region, and he considers that the nervous elements have been thrust upon and thus utilized the alimentary tube as a supporting structure. Gaskell has entered minutely into details concerning the modification induced by the change in the position of the mouth.

Which of the views is the correct one—whether the gut, becoming disused from gradual loss of function, became utilized for the support and extension of the surrounding nerve ganglia, or was rendered useless in consequence of the encroachment of nervous material, or, what seems equally probable, the change of its intrinsic elements into nerve-cells, will require further investigation.[1]

Apart from such considerations, the view that the central nervous system is disposed around a modified piece of intestine, offers an explanation of several otherwise inexplicable phenomena which will be duly considered in some of the ensuing chapters.

Tails.—Among suppressed parts, so far as man is concerned, must be included the tail. That man has descended from forms furnished with tails cannot be doubted, for at the end of the vertebral column he still carries three, four, and occasionally five rudimentary

---

[1] Those interested in this question will find the matter discussed in my monograph on Dermoids. Gaskell's views are published in the *Journal of Physiology*, April, 1889, vol. x. part 3.

DISUSE AND ITS EFFECTS. 53

caudal or coccygeal vertebræ. The most constant number in the adult is three, but in the embryo germs of five vertebræ can be distinguished. This remnant of a tail is connected with the sacrum by bands of fibrous tissue, the degenerate remnants of the muscles which in

FIG. 26.—An African child with a pendulous tumour hanging from its buttocks; false tail. (After Virchow.)

a functional tail raise, depress, or move it from side to side. This is not mere speculation, for these ligamentous bands are not infrequently replaced by muscles known as curvator coccygis, extensor coccygis, and agitator caudæ. The scepticism regarding the occurrence of tails in men is mainly attributable to the mistakes which have been made by incompetent observers in reporting, as tails, structures which had no right to such a title, and it will be useful for us to consider the various forms of true tails and the appendages which may be mistaken for them.

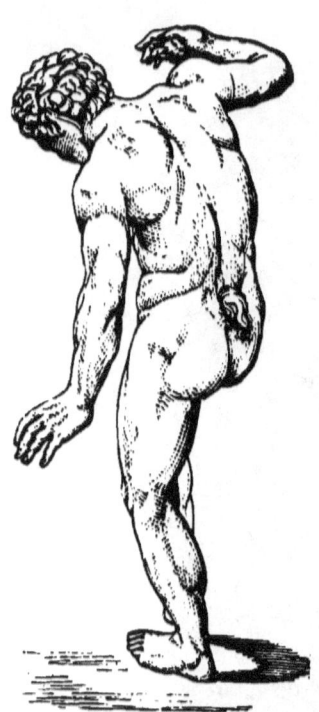

FIG. 27.—A Faun, to show the goat-like tail.

We may, with Virchow, divide tails into two classes, true and false. True tails are of two varieties: the most perfect tails are composed of bony segments directly continuous with the vertebral column, as in the case of monkeys, horses, dogs, cats, lions, &c. The less perfect variety is like that of the pig, soft and flexible. That man has descended from ancestors which possessed tails there can be little reasonable doubt, and that children are occasionally born with one and even two extra bony segments to the coccyx, as man's rudimentary tail is termed, is undoubted. I have on several occasions seen five rudimentary vertebræ in a child's coccyx.

It must also be remembered that this portion of the vertebral column may be more prominent than usual and project like a tail, yet on dissection contain but the normal number of bony elements; whilst in other cases the number may amount to five, and no abnormal projection be observable. Thus far a well-developed tail in an adult human subject containing bony elements continuing the vertebral series has yet to be detected. In the new-born child, soft tails about an inch in length have been observed: these contained cartilaginous tissue and resembled the flexible tail of pigs.

FIG. 28.—An Ægipan sporting with a Faun. Bacchus and Silenus.

Many instances of tailed children when critically examined turn out to be tumours or tufts of hair in the loin. A general notion of a false tail may be gathered from the African child represented in fig. 26. In this case a large rounded tumour hangs pendulous from the child's buttocks, and a little imagination would soon distort this into a tail. The tumour was removed in Central Africa and sent to Professor Virchow.[1] The pendulous mass consists of a hollow central cavity surrounded by fat and covered externally by skin, and in Virchow's opinion it arose as a diverticulum from the membranes of the spinal cord (*spina bifida*).

The most interesting false tails are those formed of tufts of hair. It was mentioned in the last chapter that certain malformations of the spinal column are associated with hair-fields and long tufts of hair in the loin. Sometimes, as in the example on page 23, the hairs are several inches long, recalling the goat-like tuft of hair or tail which sculptors represent in the loins of satyrs (fauns and ægipans). In fauns the tail strongly resembles the tuft of hair seen in some human beings. For instance, compare the back of the faun in fig. 27 and that of the child, fig. 12.

Virchow, in writing on this subject, points out the possibility that sculptors and artists in representing these mythical satyrs and "gods of the wood" with tufts of hair for tails, did not trust entirely to the imagination, but that such oddities had a certain amount of foundation in fact. There is much to support this view. Those sylvan deities, the ægipans, had a man's head and body, pointed ears, and the hindquarters of a goat. In some forms of local hairiness due to spinal defect, the hair extends over the legs and buttocks as in the ægipans. The cloven hoof admits of a two-fold explanation. In the first place, malposition of the foot is a frequent complication of congenital

[1] Virchow's Arch. Bd., ci. S. 571$

DISUSE AND ITS EFFECTS. 57

defects in the spine, the sole of the foot being turned upwards and inwards. In some children the middle toes may be deficient and cause the foot to assume a cloven appearance.

Recklinghausen has suggested that, as in many cases of spina bifida, disease of the bones of the foot occurs as a complication, and often induces loss of the middle toes, this may have stimulated in the imagination the notion of a cloven foot. It is perhaps not unfair to infer that from such sources as these originated the corporeal form of our much dreaded mystical devil, with hairy body, cloven feet, and tail (fig. 28).

The relation of fauns, ægipans, and goats is discussed further in chapter iv. in connection with some other structural peculiarities they share in common with goats.

Atrophy of parts when disused in consequence of injury scarcely calls for comment (though very many interesting specimens might be described) because the effects of disuse when thus induced are not inherited.

FIG. 29.—An elongated claw from a Two-toed Sloth.

As a general rule, the statement that parts when disused become reduced in size holds good, but in connection with dermal organs such as nails, horns, and claws, it is well to point out that disuse leads to enlargement. In the preceding chapter specimens were described to demonstrate the increase in the size of dermal organs when disuse and irritation from dirt, &c., were combined. A few specimens will now be considered in which disuse alone seems responsible for the

Fig. 30.—The head of a Parrot with overgrown beak.

overgrowth. The first is the foot of a two-toed sloth which lived for many years in the Zoological Gardens, London. One of its claws, or nails, whereby it hangs suspended from the branches, is very long and has almost described a circle. As far as I could learn, the matrix was in no way irritated, and the sloth lived under admirable sanitary conditions, but spent nearly the whole of its life suspended by its nails. It is well known that the beaks of parrots, when confined in cages, grow very thick and long so as to render it necessary to give them a piece of rough stone whereon they rub the beaks to keep them within reasonable proportions. An unusual beak of this kind is sketched in fig. 30. The bird was found dead in Australia, and an examination of the body did not furnish any evidence leading to the suspicion that it had ever lived in captivity. In this parrot the upper part of the beak measures sixteen centimetres following the curve ; it is difficult to imagine how the bird lived so long.

Conditions similar to this are often detected in the beaks of partridges, pheasants, peacocks, and fowls. There now remains for consideration atrophy, the result of continuous pressure ; this is of importance because it leads to interesting pathological conditions, and plays also a part of some interest in connection with the normal development of complex animals, but as this subject is beset with technicalities and requires a rather extensive acquaintance with special anatomy, it will not be discussed.

CHAPTER III.

VESTIGIAL PARTS.

UNDER the term vestigial it will be convenient to consider those parts commonly described as rudimentary, abortive, atrophied, or useless. On the whole it is better to refer to these structures as vestigial. In dealing with such examples as the denticles of the narwhal, which never cut the gum, the teeth of ornithorhynchus, or the splint-bones of horses, we have unmistakable evidence that they are remnants of structures which were functional in the ancestors of these animals. In many instances it is not easy to decide whether a diminutive, or feebly grown part, in one animal is the remnant of an organ better grown in its ancestors, or the rudiment of an organ which has arrived at a higher degree of perfection in its descendants.

This is illustrated in the hind limbs of whales. If we regard cetaceans as living representatives of land mammals which have taken to water, the hind limbs are remnants; on the other hand, if the descendants of cetaceans have acquired terrestrial habits, then, the hind limbs which do not project beyond the skin, but are deeply buried in the blubber, must be considered rudimentary or incipient structures. The whale is merely selected as an illustration of the advantage or

VESTIGIAL PARTS. 61

convenience of employing the term vestigial; there can be little doubt that its hind limbs are remnants.

The combined effects of enlargement from increased use, suppression, and change of function, has been to produce in complex organisms a large number of vestigial parts. Since Darwin considered the matter our knowledge of such parts has increased greatly, and in this chapter a few of the more important will be considered, especially those which lead to pernicious consequences.

Before entering upon this subject in detail it is necessary to make a few remarks on what are termed *useless* parts. It is very essential that care be exercised before pronouncing any part to be useless, for, as Mr. Wallace truly remarks, "much of what we suppose to be useless is due to our ignorance." It also becomes important to inquire why such supposed useless organs are perpetuated, seeing that disuse of a part tends to promote its disappearance. Darwin was of opinion that what he termed "rudimentary organs are eminently variable, and this is intelligible, as they are useless, or nearly useless, and consequently no longer subjected to natural selection. They often become wholly suppressed. When this occurs they are nevertheless liable to occasional reappearance through reversion." Subsequently Darwin seems to have changed his opinion somewhat, for in the fifth edition of the "Origin of Species" he states, in reference to adaptive changes of structure, "But I am convinced from the light gained during even the last few years that very many structures which now appear to us useless will hereafter be proved to be useful and will therefore come within the range of natural selection."

In connection with these remarks it will be well to note the very small amount of utility which will determine the persistence of an organ; take, for instance, the hind limbs of the python and viper which are only of occasional use, in connection with procreative function, yet this is sufficient to preserve them whilst all other traces of limbs have long disappeared. Mammals abound in instances of muscles which in some species are large and important, whilst in others they may subserve such trivial functions that when absent they are not missed from a utilitarian standpoint, yet even this trivial amount of service ensures their preservation. This is well illustrated by the small muscle underlying the clavicle of man, known as the subclavius. In birds it is large and powerful, raising the wing in the act of flying. In man it is small, insignificant, and steadies the clavicle during movement of the arm. It has been found as a band of fibrous tissue, and in a few cases absent.

In reference to supposed useless parts, Wallace is of opinion that the assertion of inutility in the case of any organ or peculiarity which is not a rudiment or a correlation, is not, and never can be, the statement of a fact but merely an expression of our ignorance of its purpose or origin. In the above quotation the term rudiment refers to such parts as the pineal body, the vermiform appendix, and teeth which are developed but rarely cut the gum. These are vestiges of organs probably of great importance to the ancestors of the forms in which they now persist as *reliquia*. It seems highly probable that a part which has had important functions in an animal, and then has had its function gradually abrogated by another part, is more prone to be persistent in rudiment than remnants of organs of less importance. Darwin expresses this view of the matter thus: "Organs now of trifling importance have probably been of high im-

FIG. 31.—A Horned Sheep with cervical auricles.

portance to an early progenitor, and, after being slowly perfected at a former period, have been transmitted to existing species in nearly the same state, although now of slight use." That mere disuse is insufficient to produce abolition of a part is illustrated in a striking manner by the cervical auricles in goats, pig, and man. These ears or auricles, in so far as we know, subserve no useful purpose, are extremely variable, occur in both sexes, and often are of large size (fig. 31). These auricles, as will be shown in detail subsequently, are enlarged opercula, yet an enormous space of time has elapsed since the gill-slit they guarded were functional. Nevertheless they illustrate the view I am advocating, for gill-slits and opercula were of high functional importance in the ancestors of mammalia, and are still conspicuous in the early embryo. As cervical ears or auricles will occupy our attention at some length presently, a few examples of vestigial structures, and the mischief which they now and then occasion, will be considered.

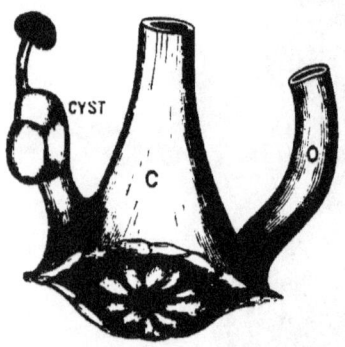

FIG. 32.—The cloaca of a Hen, showing the vestigial right oviduct. C, Cloaca ; O, Oviduct.

Hen birds possess only one functional oviduct, the left ; the chick before hatching has two, but for some reason at present obscure, the right oviduct atrophies, leaving at most a short tubular stump attached to the right side of the cloaca (fig. 32). The right ovary is either vestigial or altogether absent in birds. To the stump of the right oviduct we may very justly apply the term vestigial, and as will be demonstrated in the chapter on tumours, such vestiges are by no means devoid of danger, and even bring about the bird's destruction.

The abortion of the right oviduct in birds is in itself very curious, especially when considered in connection with the absence of the right jugular vein and carotid artery in many birds. This was attributed by Owen to the habit of birds sleeping with the head under the wing. This view is not supported by facts.

Man, in common with the four anthropomorphous apes, has attached to the lower end of the cæcum a small thin tube, which may vary from two to eight inches in length, known as the vermiform appendix (fig. 33). This tube agrees in structure with the rest of the intestine, is covered with peritoneum, possesses a muscular coat, and is lined with mucous membrane. In the early embryo it is equal in calibre to the rest of the bowel, but at a certain date it ceases to grow *pari passu* with it, and at the time of birth appears as a thin tubular appendix to the cæcum. In the newly-born child it is often absolutely as long as in the full-grown man. This precocity is always an indication that the part was of great importance to the ancestors of the human species.

Many mammals, closely allied to the anthropomorphous apes, possess very large cæca; and in some of these the terminal segment of the cæcum, although not represented as a thin, narrow tube, nevertheless resembles the vermiform appendix in that it possesses a very large proportion of the peculiar kind of tissue known as adenoid, or lymphatic.

In man the vermiform appendix is a typical example of a functionless part, and, like an idle person in a community, is not infrequently a source of considerable danger and suffering, and is responsible for a number of deaths annually. The danger may arise in three ways :—The communication with the cæcum may become obliterated, and the tube distend into a cyst in consequence of fluid accumulating within it; such a cyst may rupture, and lead to minor troubles, as local inflammation or abscess, or induce death by peritonitis. Adenoid, or lymphatic, tissue is very prone to ulcerate, and under certain conditions the adenoid tissue lining the appendix may inflame and lead to fatal perforation. A much commoner danger is the entrance into it of such things as fruit-stones and similar indigestible substances taken with the food; these act as irritants, and in the long run destroy life. A somewhat similar condition of things may be studied in lions and tigers. In these handsome animals there is no vermiform appendix, and the cæcum is even more vestigial than in man. This small cæcum occasionally contains a concretion having a fragment of bone, a nail, or piece of wood for the nucleus. Concretions in the vestigial cæcum of the tiger produce similar effects to cherry-stones, &c., in the human vermiform appendix, and on two occasions I have been able to connect such concretions with the fatal illness of tigers.

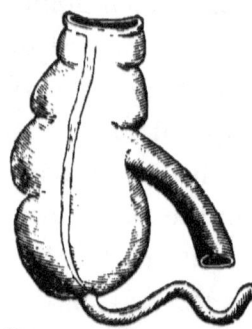

FIG. 33.—The vermiform appendix of a Silvery Gibbon (*Hylobates leuciscus*).

Although a small and insignificant cæcum is not always an advantage, it is on the other hand a disadvantage to have one too large. The horse will illustrate this: it has a cæcum measuring, on an average, one metre in length, and a capacity equal to thirty-five litres

This large cæcum is a favourite situation for intestinal concretions. These concretions are of different kinds, but they all agree in having a foreign body, such as a button, nail, hair-pin, or stray piece of metal for a nucleus, around which mineral salts in crystalline form are deposited in successive layers until the stone attains great weight—five or six kilograms is not uncommon, and a stone weighing twenty kilograms is preserved in the museum of the Royal Veterinary College, London.

Hairs from the animal's skin not infrequently form the bulk of such concretions, and may be found in the stomach of many ruminants, especially calves; but attention is mainly directed to the cæcum. The larger calculi do not give rise to such dangerous effects as the smaller, for their weight and size keep them confined to the cæcum, whereas the smaller calculi may leave this portion of the alimentary canal, and getting into narrower channels obstruct the bowel and induce death.

Apart from the special inconvenience caused by a large colon favouring the production of calculi, there is another aspect under which we may study such a question. The more remote parts of the body are from the heart the more likely are they to suffer, if from any cause the quantity of the blood in the body is diminished, or the power of the heart fails. For instance, a man suffers severely from typhoid fever, the action of the heart is weakened, and maintains the circulation with difficulty; as a result the toes, situated at the extreme limits of the circulatory system, do not receive sufficient blood, and mortify in consequence. The following is an instructive case taken from a young

Sumatran rhinoceros. This animal, like all its kind, has a large cæcum, the distal extremity of which corresponds to the vermiform appendix, and derives its blood-supply from the terminal twigs of a long ileo-colic artery. The young animal to which these remarks refer was sent to the Zoological Gardens, London; at the end of a few weeks it became sickly, refused food, and finally died. At the autopsy it was discovered that the lower jaw was extensively diseased, and a large abscess had formed. This trouble offered satisfactory explanation of the loss of appetite and inability to take food. On examining the viscera, Mr. Frederick Treves discovered the actual cause of death to be ulceration and sloughing of the extremity of the cæcum. In this rhinoceros we have a similar condition of things to the gangrene of the toes after typhoid fever, for the vital powers being reduced by the trouble in the jaw, and the inability to take sufficient food, those parts at the end of the circulatory system suffered first, and the structure most ready to succumb was the distal end of the cæcum. This case is very suggestive, because it teaches how, in the process of evolution, so far as individual parts are concerned, a limit is imposed upon the size attainable by organs, and indicates the danger to which animals are exposed in which particular vital organs attain inordinate proportions. For instance, imagine two animals living under similar conditions and upon the same kind of food, but one has a moderate cæcum, the other an inordinately large one. Should a time of scarcity or accident prevent such animals obtaining a proper supply of nutriment, the one with an average cæcum (*cæteris paribus*) has the better chance of survival.

This is merely set forth as an example of the process, and it may be illustrated by reference to parts far less vital than the intestines, viz., the antlers of deer.

The antlers when young and growing are covered with a vascular membrane, the "velvet," which bears the same relation to the antlers as periosteum bears to bone. As long as the antlers retain the velvet they live and increase in length and thickness. After a time the velvet thins, sloughs, and gradually falls from the bony portion of the antlers, which gradually dies, and, being devoid of sensation at this stage, constitutes powerful weapons of offence or defence. In time the nutrition fails in them also, and at length the dead antlers fall. The phenomenon may be explained thus: the antlers are supplied at a great disadvantage, for the blood has to travel a long distance to reach them, and is unassisted by any neighbouring anastomosing vessels such as we find in other parts of the body; consequently every inch added to the antler increases the difficulty of supply and makes its life more precarious; finally the length of the antler exceeds the distributing power of the heart, nutrition fails, the velvet is shed, and the bony tissue of the antler dies and falls. We have here conditions analogous to the cæcum of the rhinoceros, and few can doubt that those enormous antlers which decorated the head of *Megaceros hibernicus* have played a part in bringing about its failure in the great struggle of life perpetually raging in the organic world.[1]

---

[1] This view was stongly forced upon my mind during some observations made on a Wapiti deer (*Cervus canadensis*) at the Zoological Gardens. In the course of three or four months this

Teeth furnish much that is interesting in connection with vestigial structures. The enamel which constitutes a covering to the crown of teeth in many mammals may be looked upon as a vestige: no tissue resembling it occurs in any other part of the body of a mammal. Teeth are essentially calcified cutaneous papillæ; at one time in the history of our planet, her seas were peopled with numerous ichthyic forms furnished with an armour of enamel; some of the best specimens being the mailed ganoids. Under pathological conditions, however, teeth may spring up in such extraordinary situations as in cysts of the ovary.

Turning to particular cases, we may study an instructive example in the horse. This admirably specialized animal possesses three incisors and one canine on each side; then an interval follows until we reach the pre-molars: a study of closely allied fossil forms indicates that this gap, or diastema, was occupied by well-formed teeth in the ancestors of the horse, and this view finds support from the circumstance that the first pre-molar is vestigial and presents itself as a tiny socketless tooth. This functionless pre-molar is, as a rule, shed early; when persistent it is frequently a source of considerable annoyance to the animal, as every

---

animal developed antlers weighing from twenty-seven to thirty kilograms. That such rapid growth as this must tax the vital powers of an animal is clearly shown by the circumstance that during the growth of the antlers the Wapiti required, and was supplied daily with, nearly twice the quantity of food consumed by it at the time when the antlers were fully grown. The bearing of this fact from an evolutionary point of view is too obvious to need any pointing out.

veterinary surgeon knows; for horses frequently refuse food, set their coats, and get out of condition simply from the trouble caused by these teeth : as soon as they are removed the horse rapidly improves and gets once more into condition. This vestigial pre-molar of the horse is often omitted in drawings (even in veterinary works) of the teeth of the horse.

It has been clearly shown by the researches of Albrecht that man has a smaller number of teeth than he formerly possessed. The mouth is often the seat of a defect known as cleft palate ; not infrequently children affected with complete clefts are furnished with three incisor teeth in the jaw which is cleft, and occasionally on both sides. In rarer cases an extra incisor tooth may make its appearance taking rank, and being co-equal with, the normal incisors. This matter has been inquired into by many competent observers, and its occurrence is beyond all doubt. The most satisfactory explanation of the phenomenon seems to be that offered by Albrecht; it is to this effect: man normally inherits in each upper jaw germs of three incisors, one of these usually becomes suppressed ; in cases of cleft palate there is more space for the teeth to develop and a greater supply of blood to the parts adjacent ; these are circumstances favourable to the full development of the germ of the third incisor.

The fact related above is of sufficient interest in itself; it is also of importance in a general way because there is good reason for the belief that the germs of other teeth have been suppressed in the mouth of man, and that the wisdom teeth are slowly undergoing this process. It has been suggested by Mr. Eve, on very good grounds, that the germs of teeth which have been suppressed in the evolution of our species may make themselves obnoxious in an unexpected way. Teeth are formed in part by down-growths of epithelium lining the floor of the mouth; these cellular down-growths, known as enamel-organs, present distinctive features and are easily recognized by practised histologists. The

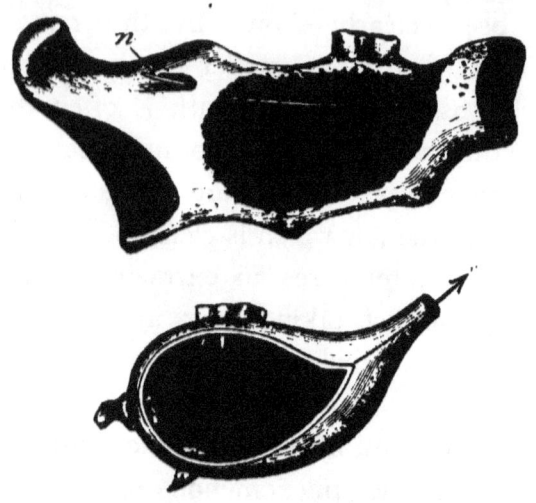

Fig. 34.—A follicular cyst developed in connection with an unerupted tooth of a Porcupine. The upper drawing shows the effects of the cyst on the jaw.

mouth of man is occasionally occupied by tumours to which the name epithelial odontomes has been given; the peculiar feature of these tumours is that they are composed of collections of cells in every way resembling the enamel-organ. The opinion receives considerable support from the fact that several carefully examined specimens of this peculiar form of tumour have occurred in cases where the number of teeth has been below the

VESTIGIAL PARTS.

normal, the tumour occupying the position which should have been filled by the missing teeth.

That imperfectly developed teeth give rise to tumours is indisputable; for instance, the teeth before they make their appearance above the gums are enclosed in a bag, or follicle, formed partly of fibrous tissue or bone. Occasionally teeth which should normally be cut and take their position in the dental series, remain hidden beneath the

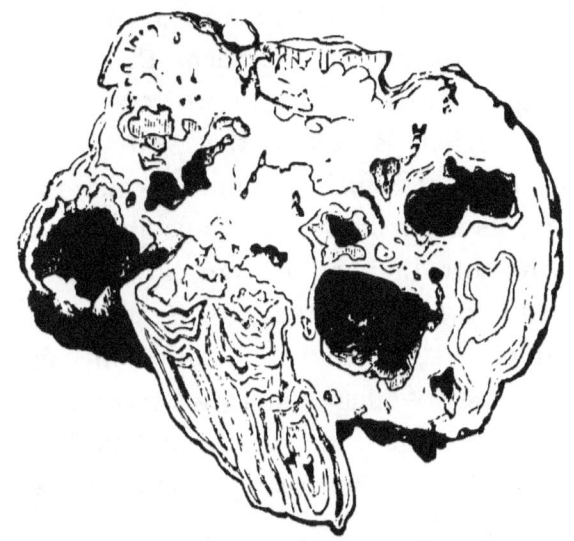

FIG. 35.—A composite odontome from a Horse, weighing .700 kilograms.

gums; in such cases fluid accumulates between the aberrant tooth and its capsules, forming a tumour known as a follicular cyst (fig. 34).

Sometimes the germs of several teeth coalesce and give rise to an ill-shaped mass of dental tissue known as a composite odontome (fig. 35): in·due course this abnormal conglomeration of tooth tissue attempts to rise above the gums, or becomes erupted like an ordinary tooth; this produces severe constitutional disturbance which may place life in jeopardy.

When dealing with the intestinal origin of the spinal cord it was mentioned that the central canal of the cord and the gut communicated with each other around the caudal end of the notochord. This connecting passage is known as the neurenteric canal, and the section of the bowel into which it opens is known as the post-anal gut, because it is situated posterior to the permanent outlet of the bowel. In the clasmobranchs this section of the primitive gut equals a third of its total length. The coccygeal region is often the seat of congenital tumours, some of which present peculiar characters. The examples most interesting to us attain a very large size—often more than a kilogram in weight—and are situated anteriorly to the coccyx. Structurally they are composed of cysts lined with epithelium, the stroma consists of very young connective-tissue: sometimes these tumours contain a portion of bowel lined with mucous membrane, possessing Lieberkühn's follicles and Peyer's patches.

A study of the development, structure, and relations of the tumours will serve to convince any impartial observer that they arise in connection with the post-anal gut; they are by no means rare; few pathological museums of any pretensions are without a specimen, or a model of them, and all surgeons of experience have encountered one or more examples of them. The large specimens are incompatible with life, but smaller ones have been successfully dealt with surgically. Finally every gradation has been recorded, from per-

VESTIGIAL PARTS.

sistence of the gut as a simple tube, with an accessory opening near the coccyx, to large growths exceeding the weight of the child unfortunate enough to possess so unwelcome an appendage.

The transformation of a piece of intestine into a central nervous system has had the effect of rendering vestigial, structures which not unfrequently behave in a manner pernicious to the individual.

For instance, the development of a spinal column to protect the cord is an outcome of this transformation, and the various defects in the development of the cord and column, if serious, are incompatible with life. These defects, known collectively as *spina bifida*, are of such frequent occurrence that in a recent careful scientific report upon this subject, it appears, that in England alone, six hundred and forty-seven deaths occurred in 1882 from this malformation, of which six hundred and fifteen were in children under one year of age. It would, in a work of this kind, be difficult to enter fully into details of the various forms of this interesting class of defects, but it may be briefly stated that many of them are failures in the formation of the bony walls, others are due to protrusions of the spinal membranes, and a rarer form arises in consequence of an accumulation of fluid inducing local dilatation of the central canal of the cord: an example of the cystic dilatation of a functionless canal.

'That remarkable appendage of the developing alimentary canal, the yolk-sac, with its vitello-intestinal duct, has already been referred to ; its significance is not easy to estimate. Although the duct connecting it with the intestine is pervious, it is not used for the transmission of yolk, for, except in the case of a few fish (fig. 36), the contents of the yolk-sac have not been detected in the alimentary canal. The yolk is taken up by the omphalo-mesenteric vessels circulating on the

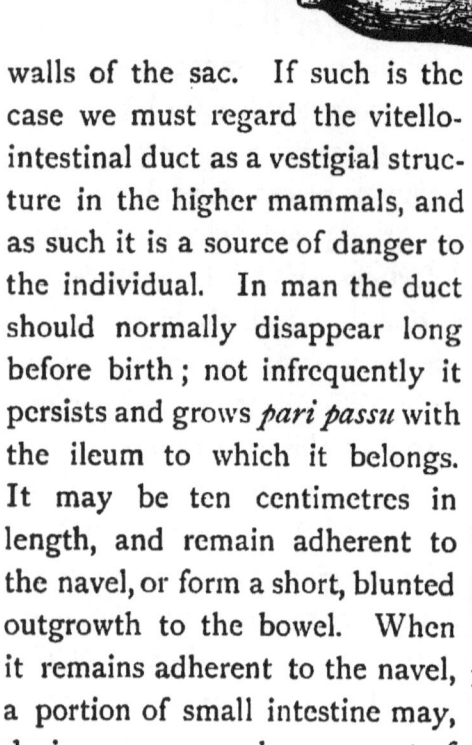

walls of the sac. If such is the case we must regard the vitello-intestinal duct as a vestigial structure in the higher mammals, and as such it is a source of danger to the individual. In man the duct should normally disappear long before birth; not infrequently it persists and grows *pari passu* with the ileum to which it belongs. It may be ten centimetres in length, and remain adherent to the navel, or form a short, blunted outgrowth to the bowel. When it remains adherent to the navel, a portion of small intestine may, during any unusual movement of

FIG. 36. — An elasmobranch fish recently hatched with its yolk-sac.

the viscera, become twisted over it, obstruction to the free passage of the contents of the bowel is established, and, unless quick relief is afforded by art, a fatal issue ensues.

The vitello-intestinal duct leads to disastrous consequences in another way. In the ordinary course of events the duct should shrivel as far as its attachment to the bowel; occasionally the process of obliteration may involve the wall of the ileum and lead to the formation of a septum, which gradually contracts and slowly causes death by obstruction of the bowel. In rarer cases the occlusion of the duct may extend to the ileum, and divide it completely. Such a condition is of course incompatible with life.

This duct is the source of other lighter troubles; those described above are the most serious.

The tongue contains a vestigial duct of great interest. At a very early period in the life-history of the mammalian embryo a diverticulum arises from the ventral wall of the pharynx, and eventually gives rise to the middle portion of that very puzzling organ—the thyroid body. For a time this duct retains its connection with the mouth; eventually the hyoid bone appears and divides the duct into two portions. The portion in relation with the mouth becomes surrounded by the developing tongue, and finally disappears, leaving nothing but a small depression on the surface of the

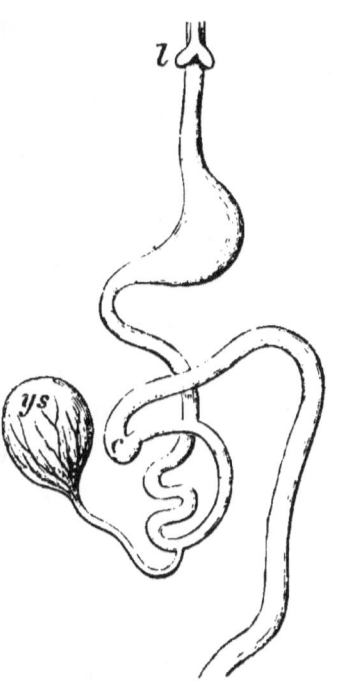

FIG. 37.—A diagram of the alimentary canal showing the yolk-sac and its duct *ys*; *c*, cæcum; *l*, lung.

organ to indicate its previous existence. A careful inquiry will reveal the occasional presence in the tongue of a duct passing from the hyoid bone to the foramen cæcum. This is not a duct which requires a microscope to distinguish it, but is capable of admitting a bristle or fine probe. In some cases this duct becomes obstructed at the upper opening, and the gradual accumulation in its interior of shed epithelium and sebaceous matter gradually distends into a large and troublesome cyst. In some cases the walls of the cyst are formed of skin, and hair may sprout from it. These cysts are not infrequent in the human subject, and have been found occupying the centre of an ox tongue, under rather unpleasant circumstances. A gentleman, whilst carving a tongue at breakfast, unexpectedly came upon a collection of hairs and fatty material in its midst, and was in no small measure astonished.

The mammalian tongue should be an organ of great interest to the morphologist; unfortunately its evolution has not yet been thoroughly unravelled. It has of course received great attention from anatomists and surgeons. From anatomical and pathological standpoints the anterior two-thirds of the tongue differ completely from the posterior third. The latter part may be regarded as the more primitive, whilst the tip of the tongue is of later development and, morphologically, less important.

CHAPTER IV.

VESTIGIAL STRUCTURES (*continued*).

THUS far a few clear examples of vestiges of organs and parts have been briefly considered; it may, perhaps, be desirable to trace some such part from as near its beginning as possible, through its various modifications and complete disappearance as a part normal to a given animal. It is also well known that many parts are present in the embryo which are not represented in the mature animal. Of this we had many instances in the preceding chapter, but such parts are often persistent as abnormalities in the adult; they are then described as being reversionary, or atavistic. As this is a subject of great interest to us, the matter to be considered now will serve as an excellent introduction to the chapter on atavism. The next few pages will be devoted to the description of cervical auricles in man and animals; the study is of interest in many ways, and especially from the circumstance that it clearly shows that disuse is, in itself, not able to bring about the complete disappearance of parts. It will also serve to illustrate the embryological rule that when parts are precociously developed in the embryo, but feebly developed in the adult, it indicates that they were of high importance in the ancestors of those particular animals.

Embryology is eloquent in furnishing evidence supporting the view that the ancestors of existing vertebrata were aquatic in their habits, that respiration was carried on in them by means of gills, and that many structural peculiarities in mammals result from the transformation of an aquatic into a terrestrial animal.

The type of respiratory organs in these ancestral forms is best preserved in elasmobranch fish, such as the dog-fish, or in a marsipobranch, like the lamprey. In such forms the water, charged with air, enters the mouth and is forced through openings in the walls of the pharynx. The pharyngeal orifices, or branchial slits, are furnished with vascular processes known as gills. In the gills, or branchiæ, the blood and water are merely separated from each other by an extremely delicate layer of tissue. Hence venous blood circulating in the gills readily gives up the excess of carbon dioxide, and as readily obtains oxygen from the surrounding water. The gills of fish and batrachians are supported upon a cartilaginous or bony framework known as the branchial bars, and in such fish as sharks a small cutaneous fold projects from each bar and covers the gill-slit as with a lid; these cutaneous lids are named, in consequence, opercula. The gill-slits, with the opercula, are sketched in fig. 36, as they are seen in a dog-fish. The first slit bears no gills in the adult fish, and is known as the spiracle, or blow-hole. In the embryo it is furnished with beautiful external delicate vascular tufts. The neck of a mammalian embryo is furnished with four similar slit-like orifices, communicating with the pharynx, as in the dog-fish, but are fewer in number. The gill-slits of birds, reptiles, and mammals differ from those of fish in that they never at any period support gills. These rudimentary gill-slits further resemble those of the dog-fish, for they present on their anterior aspect a small swelling, or tubercle, representing an operculum In the human embryo four branchial slits present themselves (fig. 38).

The first of these represents the spiracle of the shark, and in mammals become the tympano-eustachian passage, and is subservient to the sense of hearing; the small tubercles surmounting it coalesce, and gradually give rise to the pinna, or external ear, so conspicuous in nearly all land mammals. Normally the posterior gill-slits disappear. It is by no means uncommon to find in the sides of the neck of a child, along the anterior border of the sterno-mastoid muscle, small openings in the skin capable of admitting a thin probe. These congenital fistulæ, especially when they exist in the upper part of the neck, communicate with the pharynx. This in some cases may be demonstrated by allowing the child to swallow milk; drops of the milk will find their way through the fistula and appear in the neck. Stress must be placed on this simple experiment, for His, of Leipzic, has urged that branchial fistulæ in man never communicate with the pharynx, and that the connection, in those which were supposed to open into it, was the result of incautious use of the probe. This view is erroneous; I have seen milk issue from such fistulæ in individuals who have never been submitted to sounding

FIG. 38.—An early human embryo with the branchial slits.

Those occurring in the lower part of the neck end blindly. The usual situations of the four branchial slits are indicated in fig. 39.

Sometimes we find in the situations frequented by these fistulæ instead of openings small rounded white

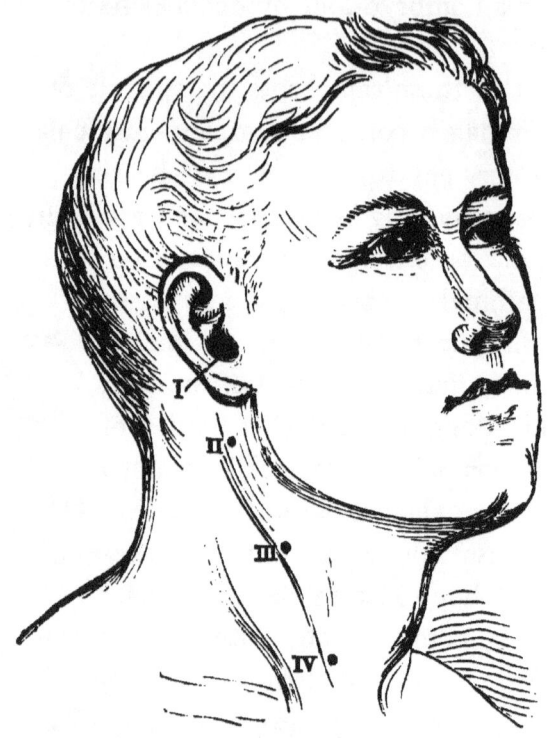

FIG. 39.—A side view of the neck; the figures II, III and IV indicate the common situations of branchial fistulæ.

patches of skin, natural cicatrices, indicating the points of obliteration of the clefts. It is by no means infrequent to find the cutaneous orifice of a persistent branchial slit surmounted by a cutaneous tag, which often contains a small nodule of yellow elastic cartilage resembling that found in the pinna; these projecting pieces of skin often

VESTIGIAL STRUCTURES.

occur unassociated with fistulæ, and are most common in two situations in the neck, at the spots marked III and IV in fig. 39. As a rule they are symmetrical; usually they are short, often looking like mere pimples on the side of the neck. In some cases they may attain a length of two or three centimetres. A very large one is represented as it grew

FIG. 40.—A Girl with a cervical ear or auricle.

from the side of a girl's neck, in fig. 40, and in a child, fig. 41. These fistulæ and cervical auricles, or ears, as they are called, usually affect many members of a family; the mother may possess cervical auricles, and one child have a cervical fistula, whilst a third may have fistulæ and auricles combined.

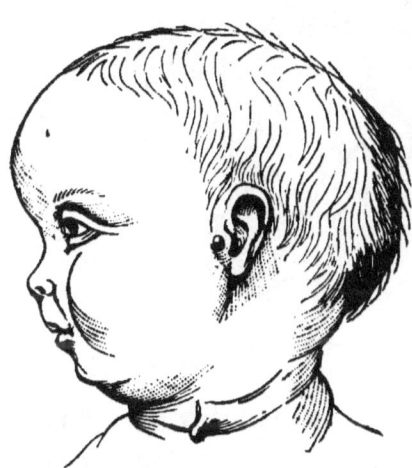

FIG. 41.—Child with cervical auricle and a supernumerary tragus on each side.

The question naturally suggests itself, if these fistulæ and auricles occur in man they should also be found in other mammals, especially as the gill-slits are as marked in them as in the human species. The inquiry is full of interest. Our knowledge of branchial fistulæ in mammals other than man is very scanty; this is not matter for surprise, as it is only of recent years that information regarding these fistulæ in him has been very exact or abundant.

FIG. 42.—A Goat with cervical auricles.

Heusinger[1] mentions the occurrence of congenital fistulæ in the horse immediately below the ear, and near the angle of the jaw. He stated that they are more frequently recognized in carriage than in draught horses; the secretion or discharge which issues from

[1] "Deutsche Zeitschrift für Thiermedicin," Bd. ii.

them soils the surrounding skin and draws attention to their existence.

Cervical auricles have been studied in the goat. In 1876 Heusinger mentioned the frequency with which pendulous tags of skin occur in the necks of pigs,

FIG. 43.—An Egyptian Goat (*Hircus thebaicus*), with cervical auricles.

goats, and sheep. Yet little has been done to support his observations. As a matter of fact, these pendulous bodies are extremely common in the necks of goats; few persons seem to notice them until their attention is particularly drawn to them.

A well-marked specimen of cervical auricle in a goat is sketched in fig. 42: these auricles are situated at a spot corresponding to the external orifice of the third branchial slit of the embryo.

The auricles are not confined to any particular species of goat; the one sketched above is a cross between a

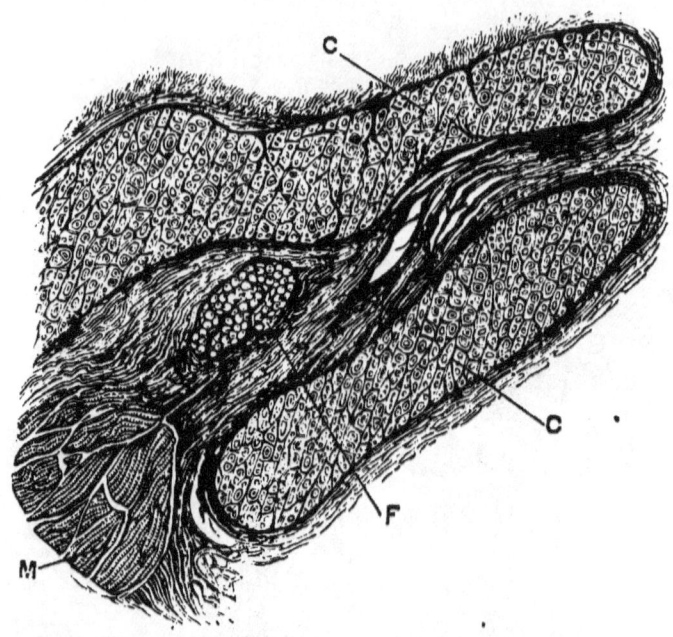

FIG. 44.—Magnified view of a section of a cervical auricle removed from a baby's neck. C, cartilage; M, muscle-fibre; F, fat.

Nubian and a common goat. In fig. 43 a sketch is given of the head of an Egyptian goat (*Hircus thebaicus*), made from life by Mr. R. E. Holding from a specimen which was living in the Jardin des Plantes, Paris. Their existence in the pig has been recorded, and a horned sheep with two well-marked cervical auricles is figured on page 63.

It has been doubted whether these pendulous bodies are of the nature of auricles, and it is desirable that the evidence should be put before the reader which favours such an interpretation. An auricle, or pinna, may be defined as an enlarged operculum in a mammal, consisting of a framework of yellow elastic cartilage covered with skin, similar to that on the rest of the body, and containing striped muscle-fibre.

The cervical auricle, such as is seen in the neck of the girl on page 83, agrees with this definition in every particular; it contains yellow elastic cartilage, is skin-covered, and has muscle-fibre attached to it, as may be seen on reference to the magnified sketch of a section of a small cervical auricle removed from a child's neck, immediately above the inner end of the clavicle. In order to complete the definition, we require to show that they are enlarged or persistent opercula.

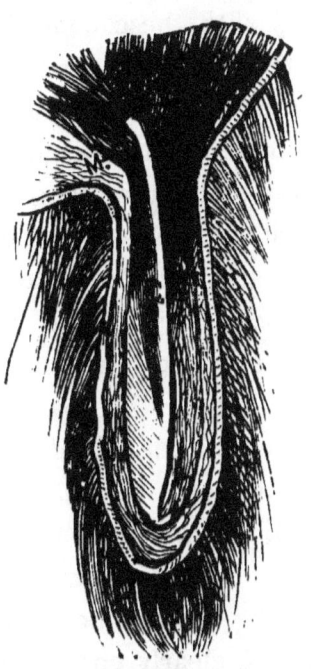

FIG. 45.—Vertical section of the cervical auricle of a Goat. C, cartilage; M, muscle-fibre. (Nat. size.)

The specimen, from which the drawing, fig. 44, was prepared was associated with a persistent branchial cleft, and in the cases where clefts are not persistent, the auricles are situated at spots exactly corresponding to the point where such fistulæ open; and, as has already been mentioned, one member of a family may have a persistent branchial cleft, another have a cervical auricle only, and a third a persistent cleft and auricle. It is also a point of some interest to remember that in the human subject the operculum of the third cleft is that most commonly seen in the adult.

The pendulous bodies in the goat harmonize admirably with these conditions. Professor Charles Stewart detected in the auricle of the goat figured on page 84, an axial rod of coarse, yellow elastic cartilage, and Franck, in his work on the "Anatomie der Hausthiere," 1883, states that in goats and pigs this rod of cartilage exists in these so-called bells (Glöckchen oder Berlocken), and draws attention to the existence also of striped muscle-fibre in them. The anatomy of a goat's cervical auricle is shown in fig. 45. As these bodies occur in goats at the situation of the external orifice of the third branchial cleft, most anatomists are of opinion that they are homologous with the cervical auricles of man.

An impartial consideration of the evidence relative to the development of the pinna in land mammals shows, clearly enough, that it is to be regarded as the confluent opercula of the first and second arch, extraordinarily developed from increased use in connection with the acoustic functions, which have gradually arisen in connection with the first branchial cleft. The remaining opercula have been suppressed partly from loss of function and partly from the excessive development of the first and second operculum. That the gradual development and increased importance of the external auditory apparatus is in a large measure, if not entirely, due to the gradual acquisition of terrestrial habits by animals originally aquatic, is largely supported by the condition of the pinna in aquatic mammalia. The adult whale has no pinna or external auditory meatus; Howes has detected vestiges of the pinna in the embryo of the white whale (*Beluga leucas*): the pinna was almost microscopic in size

FIG. 46.—The head of a Seal (*Otaria gillespii*), showing the small pinna. (After Forbes.)

and pointed, resembling in a very striking manner the small cervical auricles in man.

In most seals the pinnæ are wanting, and when present they are short, pointed, and vestigial, as in the eared-seals (*Otaridæ*), fig. 46.

If the *Cetaceans* and *Phocidæ* are to be regarded as land mammals which have taken to the water—which is the most consistent manner of studying them—we must regard the pinnæ as having slowly atrophied from disuse, but persisting as remnants in a few adult forms, and present only in the embryos of others. The persistence of the pinna in the *Otaridæ*, and the frequent presence of cervical auricles in goats are difficult to account for, especially as we are unable to assign to them any function.

The consideration of cervical auricles would be incomplete without reference to their existence on the

FIG. 47.—The head of a Satyr (Ægipan) with a sessile cervical auricle. (British Museum.)

statues of fauns and satyrs. My talented friend, Mr. S. G. Shattock, first drew my attention to this matter. They are not always represented of the same shape or size, but nearly always occupy the same situation on the neck: thus in the statue of an ægipan (satyrs with goat-like legs) in the British Museum, the auricles are sessile (fig. 47). This is the common form in man.

These appendages are not seen on the statues of

VESTIGIAL STRUCTURES. 91

modern fauns so constantly as in ancient fauns, and are usually represented as pendulous structures, admirably shown in the drawing of a faun in the Capitol (fig. 48). In all cases these pendulous skin tags in the statues are

FIG. 48.—A Faun and Goat from the Capitol, with cervical auricles.

placed along the anterior border of the sterno-mastoid muscle. It is, of course, a matter for discussion whether the sculptors obtained their notion of the cervical auricles from human models or from goats. In some cases they may have obtained them from man, but in the majority of instances, especially the pendulous auricles of fauns, the goat furnished the model. This is illustrated in the faun from the Capitol, for we see by the side of the faun a goat with cervical auricles clearly and unmistakably represented (fig. 48), and on the faun's shoulders a goat's skin is thrown. The goat element in the composition of these satyrs is evident in more ways than one ; the ægipans are goat-legged and their tails are excellent copies of that appendage in the goat. Be this as it may, we are bound to admit that the old sculptors were close observers of nature.

The grounds for regarding these congenital appendages in man as auricles may be thus summarized :—

1. Embryology teaches that they grow like the normal pinna from the swollen edge of a branchial cleft, and are thus homologous with opercula.

2. Frequently such auricles surmount the cutaneous orifice of a congenital branchial fistula.

3. When no fistulæ are present, the situation they occupy corresponds to that of the third or fourth branchial cleft. Most frequently it is the third cleft, that is, in the middle of the neck, corresponding to the anterior border of the sterno-mastoid muscle. The fourth cleft opens near the sterno-clavicular articulation.

4. Structurally they correspond to the normal pinna.

5. Not infrequently one member of a family will have persistent branchial fistulæ, whilst another has cervical auricles, and a third a fistula and cervical auricle.

Before leaving the consideration of the changes which result from the transformation of aquatic into land

VESTIGIAL STRUCTURES.

animals, it may be useful to draw attention to one condition, indirectly associated with this remarkable change, which produces greater inconvenience than can be attributed to cervical auricles. In many of the situations where canals open on a free surface, the terminal orifice of the canal is, as a rule, surrounded with glands and a collection of tissue, peculiar in structure, termed adenoid. Such glandular collections are more abundant around the terminations of functionless ducts. Some of the more characteristic examples occur in the pharynx marking the inner orifices of the branchial clefts. Of these the most conspicuous is named the tonsil.

The tonsils are familiar to all as the sub-globular shaped structures lodged in the recesses on each side of the mouth at the spot where the mouth joins the pharynx, or cavity where the nasal and buccal passages become directly continuous. The space between the mouth and pharynx is technically termed the fauces. The tonsils vary considerably in size; in some persons they are large and prominent, in others small and scarcely recognizable. Structurally they are composed of adenoid tissue, covered with mucous membrane, beset with a number of shallow crypts which secrete thick, tenacious mucus. The niche in which each tonsil is lodged is termed the tonsillar recess, and indicates the exact spot where the second branchial cleft in the embryo communicated with the pharynx; it is also the spot where the cleft, when persistent, opens internally. The connection of the tonsil and its recess with the second branchial cleft is also indicated anatomically by the glossopharyngeal nerve and the lingual artery, which, in the embryo, are distributed to this cleft.

How far the tonsil subserves any useful purpose is very doubtful: certainly they are often removed, and persons usually experience relief rather than suffer inconvenience from the loss. Of course they are only removed when enlarged from disease; and it is quite certain that the tonsils are often the seat of disease which is not merely troublesome to the individual, but is at times fraught with great danger to life.

The anatomy of this region in the horse is instructive. In this mammal veterinarians describe the tonsil as absent. In that the horse has no collections of adenoid tissue in the sides of the fauces such as exist in man, the statement is correct; but we find on each side a large cyst occupying the pharynx and constituting a chamber of communication between each eustachian tube and the nose. These large sacs are known as the guttural or eustachian pouches. A careful study of these pouches has induced me to regard them as dilatations of the pharyngeal ends of the second branchial clefts; these are the clefts from which the tonsils of man arise. It is also of some importance to remember that in the early human embryo the tonsil is represented as a sac with a slit-like opening wherewith it communicates with the pharynx. The connection of the guttural pouches with the eustachian tubes is secondary.

It is not my intention to enter in detail into the structure and relation of these curious pouches; but to point out that, like the tonsils of man, they are sources of inconvenience, trouble, and occasionally disaster. Like the tonsils, also, no known function is served by these pouches.

VESTIGIAL STRUCTURES. 95

It has already been mentioned that they communicate with the nasal chambers by slit-like orifices; consequently, when a horse sniffs, air is drawn into the pouches. Should this occur when horses are feeding in a manger or nosebag, and the food is dusty, irritant particles of dust are drawn into the guttural pouches and set up inflammation. Dust and mucus thus accumulating in the sac give rise to rounded bodies,

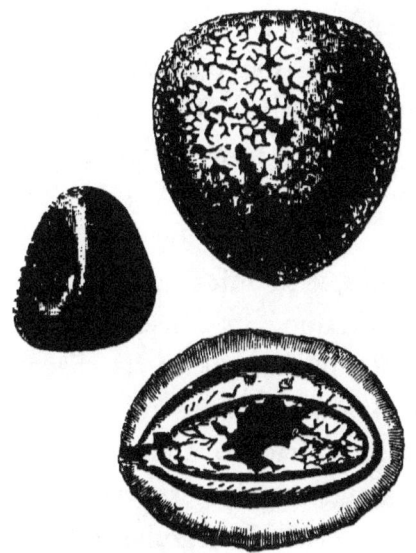

FIG. 49.—Concretions from the guttural pouches of the Horse. One is shown in section. (Natural size.)

technically known as concretions from the guttural pouches. Sometimes they are oval, sometimes bean-shaped, and vary in size from a cherry-stone to a walnut, and in number from three or four to a hundred. Their consistence resembles cheese, and on section exhibit a laminated structure (fig. 49).

When the inflammation is more intense pus forms, requiring active interference for its relief and cure. There is reason to believe that millers' horses are more liable to acquire these concretions than others.

There is a third tonsil in the pharynx which deserves some attention, although it is not associated with a branchial cleft.

When discussing the probable mode of origin of the central nervous system, attention was drawn to a duct which, in the embryo, traverses the floor of a recess in the base of the skull (the pituitary fossa) and opens on the roof of the pharynx. This duct is represented in fig. 25, page 50, and is now regarded as a remnant of the ancestral vertebrate gullet. The pharyngeal orifice of this duct is surrounded by a collection of tissue which, structurally, is identical with the tonsils, and the organ has been named in consequence the pharyngeal tonsil of Luschka, in honour of its discoverer. As far as is at present known this organ has no function, but it is often a source of trouble and inconvenience, mainly in children, inasmuch as it is especially prone to enlarge and obstruct the eustachian tube, producing deafness. It also interferes, when large, with the free passage of air through the nostrils, to such a degree as to require surgical interference.

As most of the vestigial structures considered in this chapter are the outcome of modifications induced by the change from an aquatic to a terrestrial mode of life, we may conclude it by briefly considering the fibula in relation to Pott's fracture. No bone of the lower limb of man, excepting the neck of the femur, is so liable to fracture as the fibula in its lower fourth. This accident is attended with certain peculiarities, and is named after the great surgeon Percival Pott, who first accurately described them. The chief features of Pott's fracture are the following: The fibula, or small bone of the leg, is broken about seven centimetres above the ankle, the tibial malleolus is splintered off, or the deltoid ligament ruptured, and the foot everted. The most frequent cause of this very common accident is a sudden and violent twist of the foot. In order to study thoroughly the conditions which predispose to this accident it will be necessary to briefly review the history of the fibula, and it is a fact of some interest that no one has ever described the occurrence of Pott's fracture in any mammal save man.

An examination of the hind limb of a menobranchus, or menopoma, will serve to show that the bones of the leg—the tibia and fibula—are equal in size. In such animals the legs are used chiefly as paddles, enabling them to move freely in water. The descendants of some of these forms changed their mode of life, becoming semi-aquatic, or entirely terrestrial animals, and began to use their limbs for creeping, crawling, or running habits which led to changes in the bony framework. In the case of the leg it is easy to see that it is advantageous for the weight of the body to be transmitted to the ground by one bone rather than two, hence the bone most used increased in size: this enlargement would induce a deviation of blood in favour of the bone most used—the tibia—to the detriment of the companion bone—the fibula. So truly does the fibula obey the law of heredity, that in the embryo it to some extent maintains its pristine eminence. This is strikingly shown in birds. In the chick at the fifth day of incubation the fibula equals in length, and nearly in thickness, the tibia. Subsequently it dwindles, and in the adult bird it is represented as a slender style of bone appended to the proximal extremity of the tibia. In man the tibia, as compared with the fibula by weight, is as three to one: at the third month of embryonic life the fibula has a transverse section nearly equal to that of the tibia. Even in adult life if the tibia be broken and fail to unite, extra work is thrown upon the fibula, and in course of time this bone will enlarge, and its shaft, as I have been able to demonstrate, may exceed in thickness that of the tibia. Darwin refers to some experiments of Sedillot in which small portions of the shaft of the tibia were removed in young dogs: the result was that the fibula, which in dogs is almost as slender as in birds, became greatly increased in size consequent upon the extra work required of it. As additional evidence in support of the view that the small size of the fibula in comparison with the tibia is due, indirectly, to the change of function of the leg from a paddle to an organ for land locomotion, it may be mentioned that in such aquatic mammals as seals the fibula is not so small in proportion to the tibia as is the case with terrestrial mammals. It is on these grounds that we may reasonably believe that the small size of the fibula, in comparison with the tibia, may be included as one of the changes resulting indirectly from the gradual change of an aquatic into a terrestrial animal.

VESTIGIAL STRUCTURES. 99

As the fibula in nearly all mammals is thin, slender, and almost vestigial, we have yet to inquire how it is that Pott's fracture is peculiar to human beings. A comparison of the human malleoli with those of mammals shows that man differs from them in that the external or fibular malleolus descends much lower than the tibial malleolus: even in those mammals which so closely approach man in anatomical characters as the gorilla, chimpanzee, orang, gibbon, and macaque, the malleoli are on the same level. In the accompanying sketches, fig. 50, the

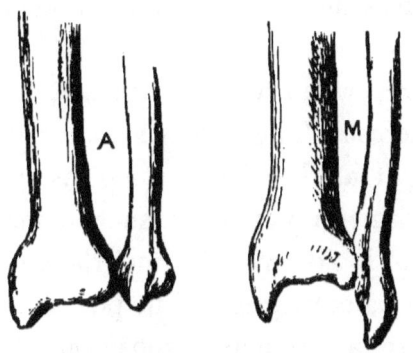

FIG. 50.—A, the malleoli of the Chimpanzee; M, the malleoli of Man.

malleoli of man and a chimpanzee are introduced for comparison.

In 1886 Gegenbaur published the highly interesting observation that, in the human embryo at the fifth month of intra-uterine life, the tibial is more prominent than the fibular malleolus; at the seventh month they are equal; from this date onward the fibular exceeds in length the tibial malleolus. Thus at the fifth month the human malleoli present a condition common to the majority of mammals; at the seventh month they correspond to the simian type, and subsequently assume the relation normal only in man. This extra length of the fibular malleolus gives great firmness to the ankle-joint, and has probably been acquired concurrently with the assumption of the erect posture. It is to this extra length of the outer malleolus, associated with the slenderness of the fibula, that the frequency of Pott's fracture may be largely attributed; the long fibular malleolus affording good leverage when the foot is violently and suddenly twisted laterally, the force applied to the long distal end causes the fibula to snap at some point in its lower fourth.

This inquiry, when pushed further, leads to other points of interest. The foot of an orang, instead of forming a right angle with the leg, as in man, has its inner border drawn upwards in such a manner that the sole of the foot looks inwards, and the back, or dorsum, of the foot looks outwards. This position of the foot is associated with a peculiar disposition of the articular surfaces of the astragalus, or ankle-bone. In the human embryo, up to the seventh month, the foot has a similar position, and the articular surfaces are disposed as in the orang; after the seventh month the foot gradually passes into the position characteristic of the adult, but not infrequently it retains the simian position and the child is said to be club-footed, or, properly speaking, it has talipes equino-varus. Messrs. Parker and Shattock have clearly shown that the articular surfaces of the astragalus in cases of congenital talipes equino-varus retain the ape-like disposition, and it occurred to me that if this is constant, the fibular malleolus in children with congenital talipes equino-varus should also retain the simian type and not exceed in length the tibial malleolus. Careful dissections of such limbs have shown this to be the case, and in specimens of congenital talipes equino-varus in children just born, and in those who in spite of the deformity have attained mature age, the tibial and fibular malleoli are equal in length.

Summarizing these facts, we find that in the human embryo the fibula gives evidence that primitively it was nearly, if not quite, equal to the tibia in size; that during development the malleoli present, in length and relation to the astragalus, conditions which are permanent in mammals closely allied to man; that the increased length of the fibular malleolus firmly fixes the foot in the standing position; and that the thinness of the fibula and its long malleolus is accompanied by the inconvenience of predisposing him to the occurrence of Pott's fracture.

CHAPTER V.

DICHOTOMY.

A STRONG tendency exists in the animal and vegetable kingdoms for parts ending in free extremities to bifurcate or dichotomize. In many instances partial or complete dichotomy occurs so constantly that it is regarded as a normal condition. Many extraordinary and beautiful forms among animals depend upon its occurrence, as well as a large number of malformations when dichotomy occurs abnormally.

The principle may be illustrated by the star-fish. In many specimens of this invertebrate we find normally five arms arranged radially around a central disc. It is not uncommon to find, as in fig. 51 (A), an arm reduplicated—this is complete dichotomy, and as the two halves of the bifurcated ray are symmetrical we speak of it as equal dichotomy. On the other hand, the ray may be double only at its extremity. In fig. 51 (B) is a star-fish ray drawn upon a larger scale in which the dichotomy is only partial.

Every supernumerary ray attached to star-fish is not due to dichotomy. When an arm is lost by accident and the corresponding segment of the disc is uninjured, or only slightly damaged, one or more rudimentary rays may grow from it. Such rays are produced by a process termed "budding." Very little experience enables an observer to distinguish between a dichotomized ray and a bud.

In 1831 Tiedemann [1] published a brief description of a star-fish, *Asterias equestris* Linn, with a partially dichotomized ray, and pointed out that it must be regarded as

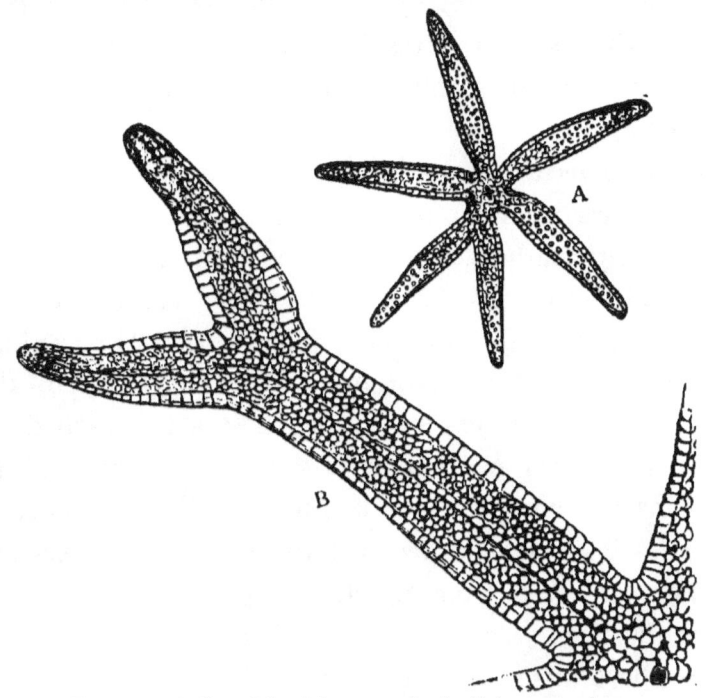

FIG. 51.—A, Star-fish with a completely dichotomized ray; B, a larger ray partially dichotomized.

a malformation and not regeneration, as is so frequently the case with star-fish.

Dichotomy occurs also in the appendages of the skin. Few parts seem more distinct than feathers and teeth, yet a study of the manner in which these

[1] "Zeitschrift für Physiologie," Bd. iv. p. 121.

structures are developed, demonstrates conclusively enough that feathers, hair, and teeth are specialized papillæ of the skin. It is not my intention to discuss this question, especially as the researches of anatomists have long established the truth of this generalization. As a rule feathers and teeth are formed from simple papillæ, but occasionally a papilla will dichotomize; should this occur, the result is a bifurcated feather or, in the case of the dental papilla, two teeth will appear fused together.

Most of us are acquainted only with single feathers, such as are used for making pens. If, instead of the large wing feathers, we select one from those covering the body—contour feathers, as they are called—we find in very many birds each quill bearing two vexilla; the second is called the *aftershaft* or *hyporachis*, the part of the feather by which they are attached to each other is the *calamus* or *quill*. In some birds, such as the emu, the feather and aftershaft equal each other. The two forms of aftershaft are represented in fig. 52. A is from the Himalayan Monaul, and B from the Emu (*Dromæus novæ-hollandiæ*). These forms of feathers arise from dichotomy of the feather-papilla. It is difficult, without specially investigating the matter, to be sure whether the emu's equal-sized feathers are due to equal dichotomy of the papilla, and that of the monaul to unequal dichotomy, or if, in the last case, the feather grows at a greater rate than the aftershaft and stunts it.

Hairs grow from cutaneous papillæ in the same way as feathers; occasionally in hairy men hairs are fur-

DICHOTOMY.

nished with an aftershaft exactly resembling, in its relation to the main hair-shaft, that figured in fig. B. In the case of teeth it is not unusual in man to

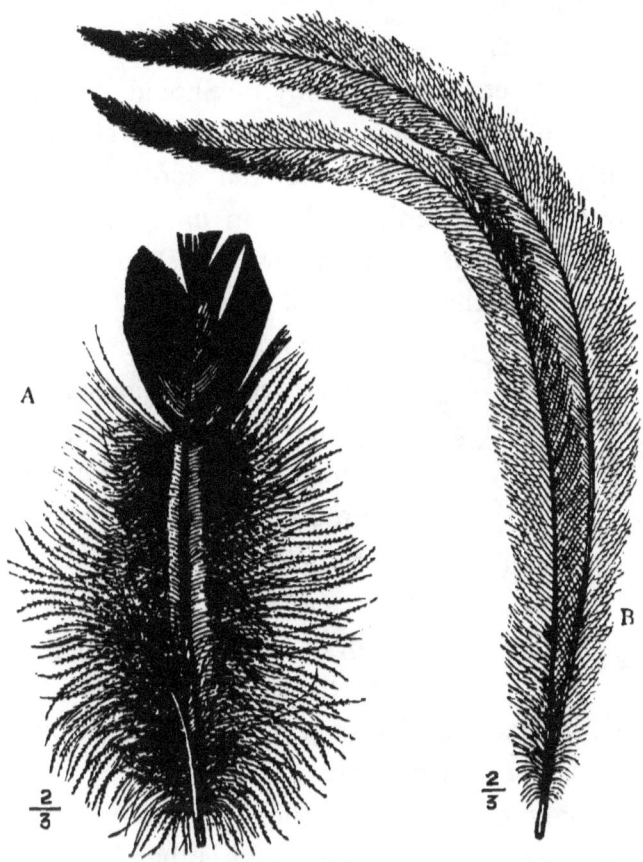

FIG. 52.—A, feather and aftershaft of the Himalayan Monaul (*Lophophorus impeyanus*) ; B, feather and aftershaft of the Emu (*Dromæus novæ-hollandiæ*).

find a double incisor or bicuspid tooth. In such cases the fangs and crowns are usually firmly united together, but the line of union is indicated by a deep, well-pronounced furrow. Such a condition, of the teeth is termed by odontologists, gemination (fig. 53). This abnormality is not confined to the teeth of man, but has been detected in those of other mammals, wild and domesticated.

Typical cases of gemination may be considered as examples of equal dichotomy. Should the process only involve a part of the papilla, we should then have only an extra crown to the tooth if it affect the incisors or a canine; but in unequal dichotomy affecting the germ of a molar we should probably find it increase the number of the cusps.

FIG. 53.—Two geminated teeth, illustrating equal and unequal dichotomy.

Antlers of deer occasionally furnish instructive specimens. Dichotomy may affect the beam or the tine producing puzzling deviations. The museum of the Royal College of Surgeons contains a pair of antlers of the moose (*Alces machlis*), in which the broad palm, so characteristic of this deer, is reduplicated in each antler (fig. 54). The specimen is Hunterian, and said to come from America.

Dichotomy occurs in the limbs of many vertebrata from the lowest to the highest, and is the chief cause of supernumerary fingers and toes, and many examples of reduplicated limbs. Many simple and uncomplicated cases of dichotomy are presented by the digits: the simplest are those in which children are born with a small fleshy nodule hanging from the finger, usually the fifth. As a rule the corresponding digit of each hand is affected. The nodule consists usually of a terminal phalanx, furnished with a rudimentary nail, attached by a slender pedicle to a well-formed finger. These rudimentary digits may be attached to the side of the terminal phalanx, but sometimes swing from the side of the first or proximal phalanx. It was this form of supernumerary digit which Darwin erroneously believed was reproduced after removal. Dichotomy of

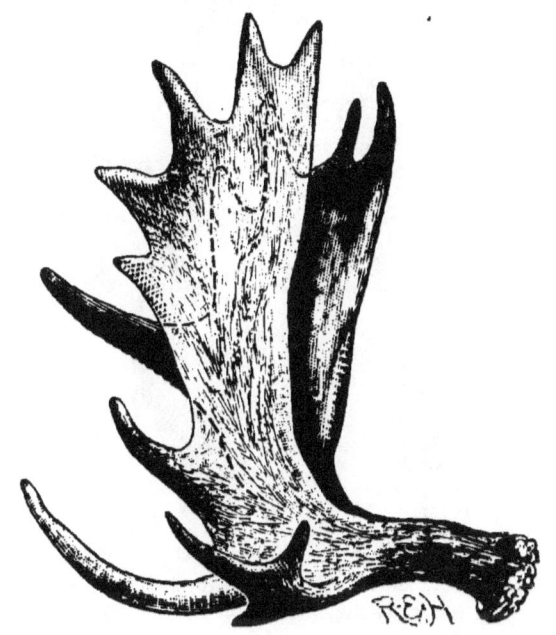

FIG. 54.—The right antler of a Moose with reduplicated palm.

the finger may be equal but incomplete; it then gives rise to the malformation shown by the thumb (fig. 55). When the process is complete, two perfectly distinct thumbs result. Such reduplication is not necessarily confined to a finger or a thumb, but may affect several digits of the same hand or foot. Jonathan, son of Shimeah the brother of David, slew at Gath a man of great stature "that had on every hand six fingers and on every foot six toes, four and twenty in number."[1] Many such cases have been observed in modern times. Supernumerary toes and fingers run in families not merely in man but in cats and dogs. Instances are

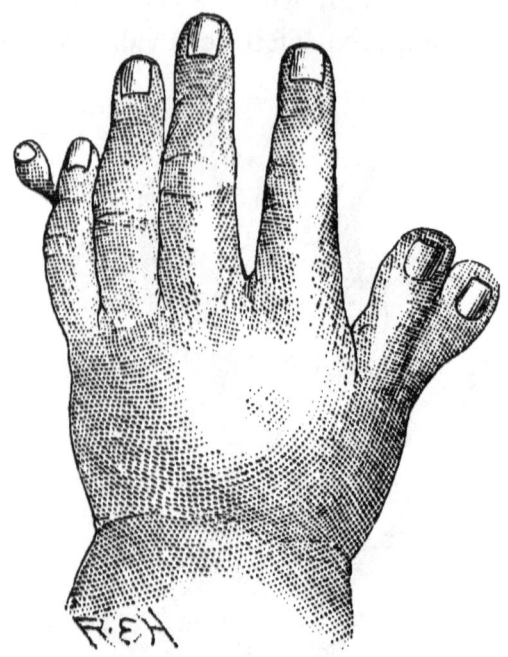

FIG. 55.—The hand of a baby, showing two forms of supernumerary digits (semi-diagrammatic).

known in which cats have six-toed kittens as regularly as Dorking fowls present five pedal digits.

In some specimens the dichotomy extends beyond the finger and involves the metacarpal bone. This is shown in the case of the silvery gibbon (fig. 56): the fifth digit was reduplicated and the distal end of the

[1] 2 Samuel xxi. 20, 21.

fifth metacarpal bone. The opposite hand was similarly affected.

When dichotomy extends beyond the fingers and metacarpal bones it may involve the terminal segment and lead to reduplication of the hand. Accessory hands or feet due to dichotomy are, in man, of very great rarity. An excellent specimen of double hand has been described by the late Dr. Jardine-Murray.[1] A sketch of the hand is given on next page (fig. 57). It may easily be conceived that should dichotomy involve a greater extent of the axis of the limb, we should get an accessory arm or leg. Accessory limbs arising in this way are, in man, very rare; indeed no such specimen is known to me, but it undoubtedly occurs in other vertebrata. Supernumerary legs are met with in the human subject, but, as will be seen later, these arise from dichotomy of the trunk axis. It should be mentioned that all specimens of polydactyly do not arise from dichotomy. Some are atavistic:

FIG. 56.—The left hand of a Silvery Gibbon (*Hylobates leuciscus*), with dichotomy of the fifth finger and distal segment of the metacarpal bone.

atavistic polydactyly can only occur in non-pentadactyl mammals, but even in them, as will be shown in the chapter on Atavism, the number of the digits may be increased by dichotomy.

There is reason to believe that supernumerary digits and limbs may be produced by cleavage throughout the vertebrate sub-kingdom. Albrecht has figured a mud-fish (*Protopterus annectans*), with bifurcation of the right pectoral limb (fig. 58). The specimen is preserved in the museum at Konigsberg.

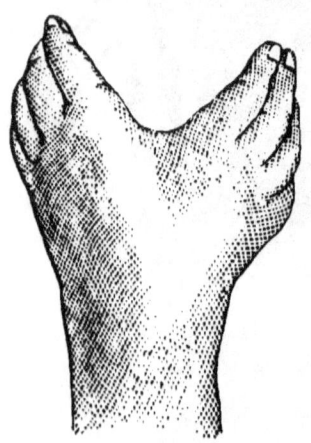

Fig. 57.—A double hand. (After Jardine-Murray.)

Haswell and Howes have shown and admirably illustrated the tendency manifested by the fins of Ceratodus to bifurcate. This is seen not only in the radii supporting the fringe of horny rays, but in the main or supporting axis.[1]

Among amphibians supernumerary limbs are not rare. An example from the common toad is sketched in fig. 59. The axis of the limb is at right angles to the trunk, and articulates with the ilium by a perfect ball-and-socket joint. The additional limb has the usual segments of femur, crus, and pes. The muscles were well-developed.

This specimen illustrates a condition of frequent occurrence in such limbs; it is furnished with an abnormal number of digits—the usual number is five, whereas in this case the pes presents seven.

[1] "Proc. Zool. Society," 1887.

DICHOTOMY. 111

Excess in the number of the limbs may occur with the fore-limbs. The museum of the Royal College of Surgeons possesses a specimen of a frog, thus described in the catalogue:—

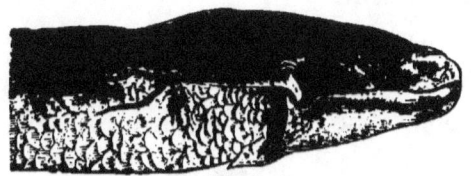

FIG. 58.—The Mud-fish (*Protopterus annectans*), with partial dichotomy of the right pectoral limb. (After Albrecht.)

" A frog with a small additional anterior extremity springing from the posterior and lateral part of the sternum " (Hunterian).

FIG. 59.—A Toad (*Bufo vulgaris juv.*), with supernumerary hind limb.

A specimen which came under my observation is sketched in fig. 60. The limb furnished with four digits was attached to the anterior part of the left moiety of the shoulder girdle through the intervention of a bone resembling a supernumerary coracoid. To judge from published cases it would seem that supernumerary limbs in amphibia are uncommon, but inquiry satisfies me that they are, in frogs and toads at least, by no means infrequent.

We will turn from amphibians to birds. In these highly specialized and exquisite forms, dichotomy is very common. The Dorking fowl has long attracted

FIG. 60.—A Frog (*Rana temporaria*), with a supernumerary fore-limb.

attention in that it presents almost constantly a double digit on the pes, increasing the number to five. This extra toe is due to dichotomy of the digit attached to the vestigial first metatarsal; it possesses three phalanges, and is furnished with a claw. This deviation from the number normal among fowls is further interesting in that it is transmitted truly to offspring when the Dorking is crossed with breeds furnished with four toes.

Dichotomy may, not infrequently, affect other digits; thus in fig. 61 a chick is represented with two bifid toes and partial duplicity of the left leg. Birds with these accessory parts may live and attain the adult condition.

Supernumerary legs are very common in birds, indeed

Fig. 61.—A Chick with two dichotomized digits and a supernumerary leg.

almost every poultry breeder has seen examples, yet accessory wings are very rare. One specimen only is known to me, a dove, preserved in the museum of the Royal College of Surgeons (fig. 62). This bird has an accessory wing growing from the lower part of the sternum.

The apparent rarity of supernumerary fore limbs, in comparison with hind limbs, will be dealt with when considering the question of dichotomy as manifested in the trunk.

In cats and dogs dichotomy of the terminal segments of the fore and hind-limbs is fairly often seen, and a five-legged dog is one of the usual exhibits at a penny

Fig. 62.—A Dove with an accessory wing, probably due to dichotomy.

monstrosity-show. An uncommon example is represented in fig. 63; it is a sheep with an extra fore-limb attached to the scapula. The terminal segment is reduplicated. Dissection of the parts seemed to show that the abnormal limb was due to dichotomy of the limb axis, but this explanation is not altogether satisfactory.

From supernumerary limbs we may now pass to the consideration of what are commonly known as double

DICHOTOMY. 115

monsters, and endeavour to show that they arise from dichotomy, partial or complete, of the trunk-axis of the embryo. Before describing actual specimens it will be well to adduce facts in support of the view that double embryos may arise from a single ovum.

In 1869 Haeckel [1] showed that it was possible, by

FIG. 63.—A Sheep with supernumerary fore-leg.

dividing artificially the eggs of *Crystallodes rigidum* to multiply the number of embryos. The experiments were very simple, and consisted in taking a Crystallodes larva on the second day, in which the amœboid movements were very lively, placing them in a watch-glass with sea-water, and with the aid of a simple micro-

---

[1] " Zur Entwicklungsgeschichte des Siphonophoren." Utrecht, 1869.

scope and cataract needle dividing them into two, three, or four pieces. Detailed accounts of six experiments are given in which the ova were simply divided, cut into three portions, or quartered. The results of these experiments were as follows :—

1. Development continued in the divided pieces.
2. The smaller the piece the slower the growth of the larva.
3. The smaller pieces tended to form incomplete individuals and inclined towards monstrosity.

These observations are valuable, for it must be borne in mind that the growth of a morula (segmented ovum), after artificial division, differs very much from the formation of a hydra out of a piece cut from an adult hydra.

From *Syphonophora* we may pass to worms. In 1828 Dugès[1] presented to the Académie Royale des Sciences, a paper entitled, " Recherches sur la Circulation, la Respiration et la Reproduction des Annélides à Branches," which contains the following remarks relative to the eggs of the worm *Lumbricus trapezoides* :—

"The first of these eggs which I opened embarrassed me much. I saw escape with a glairy material a living, white, soft, transversely wrinkled, vermiform animal, composed of a body terminated by two appendages marked from right to left by a regular spiral. It was a monster formed of two individuals joined together, fused in a part of their length, as I have since observed in others but with less symmetrical conformation. In each egg I have constantly found plunged in the same albuminous jelly, either two germs, two cicatricula, or

[1] "Annales des Sciences Naturelles," tome xv. p. 248.

two embryos, except that one of the two germs, though not aborted, merely left traces of its former existence."

In a foot-note Dugés makes a further observation: " Even in the ovary we perceive that these eggs present two distinct cicatricula, in some isolated, in others

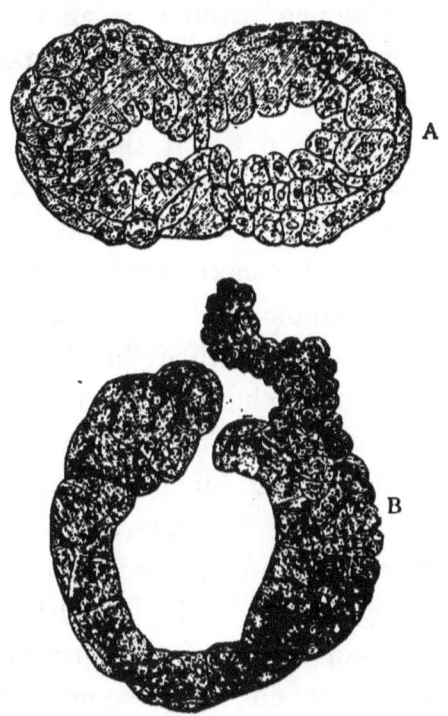

FIG. 64.—A, Transverse section of a double embryo of *Lumbricus trapezoides*.
B, Transverse section of an embryo of *L. trapezoides*. One embryo is suppressed and appears merely as an excrescence. (After Kleinenberg.)

contiguous." Dugés illustrates the phenomena he describes by some crude drawings, but there can be no ambiguity about the facts to which he drew attention.

The embryology of the worms concerning which Dugés made the above curious observations has been carefully studied by Kleinenberg,[1] and he has succeeded in tracing the development of these worms step by step, and shows beyond any doubt that it is a normal condition in *L. trapezoides* for a single germ to produce two embryos. For a time the embryos are united and turn gently in the albumen without at all impeding each other. The commissure uniting them relaxes gradually, then breaks, and the embryos are freed. There are cases not at all rare in which this singular form of development leads to monstrosity. In fact, among perfectly developed worms, double monsters occur presenting all grades of concrescence, from those firmly united along the whole extent of the body, so that separation is impossible without breaking the embryo into pieces, to others which are hatched coupled together, but only by a thin, frail ligament that the worms easily effect a separation. These junctions are always confined to the epithelial layer of the body wall.

Kleinenberg states that he has never found one of these eggs giving rise to a single embryo. It is true that a single worm escapes from a capsule, but then nearly always the remains of its companion are found. The accompanying sketches exhibit the extreme forms of these double embryos. In fig. 64 A, they are shown when of equal size; in B, one of the embryos has undergone suppression.

We have direct evidence among vertebrata that two embryos may arise from a single ovum. It has been actually witnessed in a batrachian by Clarke.[2] In the

[1] *Quart. Journal of Micros. Science*, vol. xix. 1879.
[2] "Ann. Mem. Boston Soc. Nat. Hist." 1880.

spring of 1879 he had in an aquarium two or three thousand eggs of *Amblystoma punctatum* for the purpose of studying their development. One day he chanced to find one with the medullary folds nearly completed, but they had not united at the cephalic end, and appeared to be much rounded at their anterior ends, instead of having the ordinary vague outlines; he kept it apart, therefore,

FIG. 65.—A two-headed Foal; anterior dichotomy.

and watched it. Each free portion of the medullary fold developed a perfect head, which, at first partly united, became gradually more so, until they were connected throughout their entire length. Posterior to the heads there was no sign of duplicity.

In this case a two-headed monster, with a regular symmetrical body, was developed from one egg, and the anterior portion of each medullary fold gave rise to a head.

Double embryos vary greatly, according to the degree of dichotomy, and the subsequent growth of each half of the ovum. The cleavage may affect the cephalic extremity only; this is conveniently called anterior dichotomy. Of this the two-headed foal sketched in fig. 65 will serve as an example.

In many cases the cleavage only involves the facial portion of the skull, thus producing an animal with two tongues and four pair of jaws. The supernumerary jaws are, in such cases, conjoined into a single mass wedged in between the functional jaws, and not infrequently mistaken for congenital tumours. We shall return subsequently to malformations of this class.

When the cleavage is more extensive than in the case of the foal it may give rise not only to two heads and necks, but the thoracic region of the body is reduplicated. And thus we have a single pair of legs, a common pelvis, but the anterior part of the embryo double. Several such cases have been recorded in mammals, even in the human subject, some of which have been made objects of careful physiological study, and subsequently carefully anatomized. The most important case of this kind was the celebrated Ritta-Christina, born at Sassari, in Sardinia, 1829, who after surviving the birth eight months and a half, died in Paris. Isidore Geoffroy Saint-Hilaire [1] gives an interesting *résumé* of the chief features of this case.

When dichotomy is more complete, two individuals

[1] "L'Anomalies des Organisation," tome iii. p. 119.

DICHOTOMY.

attached only by the pelvis may result, every part of the body and limbs being reduplicated, as in the case of the Two-headed Nightingale.

Such specimens occur throughout the vertebrate sub-kingdoms. Two embryo sharks are shown in fig. 66. They are united in the ventral aspect in the caudal region; the remains of a single yolk sac exists between the pectoral fins, and serves as additional evidence to indicate their origin from a single yolk. These sharks, with some similar specimens, are preserved in the museum of the Royal College of Surgeons. The catalogue states that "a female shark was taken in the Indian Ocean. When brought on deck and cut up, about thirty young escaped from the abdomen. The specimen lived for two days in a bucket of sea-water. Several examples of sharks, trout, mackerel, and salmon reduplicated in this manner have been recorded. Rauber has contributed some excellent observations on this subject. He has been successful in detecting many cases of duplicity of the medullary folds, and the evidence seems to indicate that, had they continued to develop, double embryos would have resulted.

FIG. 66.—Two embryo Sharks joined ventrally.

From examples in which dichotomy gives rise to a double-headed monster as in the snake (fig. 67) or the foal, to more complete forms, such as Ritta-Christina, the Two-headed Nightingale, or the sharks (fig. 66), we pass on to instances in which the bond of union is merely a narrow fleshy band, as in the Siamese Twins. From these it is but a step to the origin of separate twin-fœtuses by dichotomy of a single ovum. In all cases of duplex monsters which have come under my notice, the individuals composing a double monster were of the same sex, and there is good ground for the belief that when twins are of the same sex and enclosed in the same membranes they are the product of a single ovum.

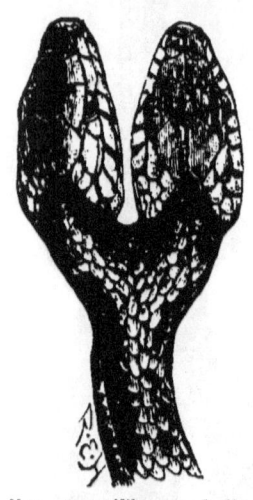

Fig. 67.—The cephalic extremity of a double-headed Snake.

Up to this point we have been considering duplex forms in which the body on opposite sides of the cleft equal each other in development. We are now in a position to consider specimens in which the dichotomy is unequal, or if equal at the outset one half grows at a less rate or becomes in great part suppressed. Such a case is shown in the calf (fig. 68). Here we have attached to the sternum of the healthy calf the headless trunk and limbs of a second calf. To such specimens the term *parasitic* fœtus is usually applied, whilst the normal calf is called the *autosite*.

Several specimens of this kind have been investigated in the human subject. A well-known case is that of the Chinese lad, Ake. He had, like this calf, the trunk and limbs of a second individual attached to the sternum.

An instructive individual similar to Ake is the Hindoo

FIG. 68.—A Calf with a parasitic fœtus attached to its sternum.

lad, Lalloo, recently exhibited in London and various parts of England. Lalloo is a highly intelligent lad of seventeen years, and has attached to the sternum near the xiphoid cartilage the trunk and limbs of a second male individual. When blindfolded he was able to accurately localize pricks from a pin made upon the parasite. Although the parasite seemed to possess independent reflex centre, it had no power of spontaneously moving the limbs, neither had the autosite any control over them.

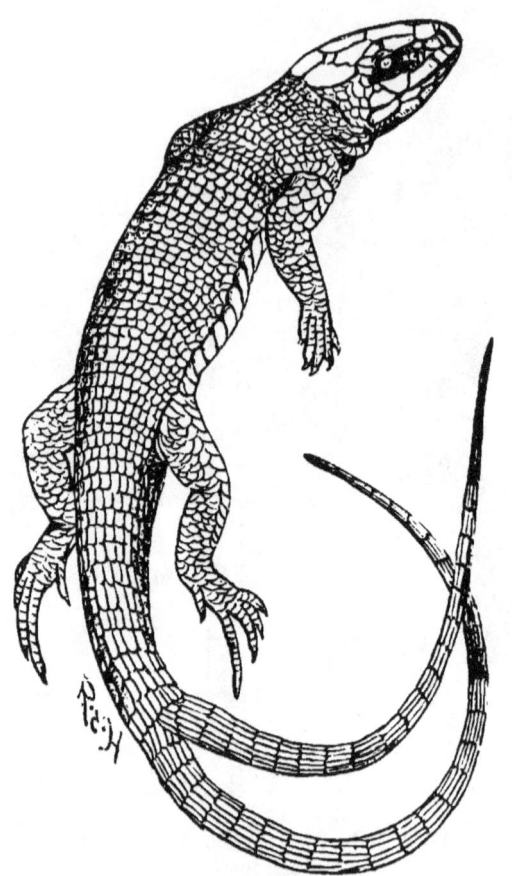

Fig. 69.—A Green Lizard (*Lacerta viridis*), with dichotomy of the tail.

Parasitic fœtuses similar to Ake and Lalloo and the calf occur in cats, dogs, and lambs. In such cases the nature of the abnormal appendage is obvious enough, but in cases where the growth of the second embryo is still further arrested, with nothing but a confused conglomeration of tissues and organs, without any definite shape to guide the observer, there is often more difficulty in coming to a conclusion. Shapeless masses of this character containing elements of a second embryo in the form of a tumour consisting of bone, liver, teeth, intestines, digits, and the like, are known as teratomata. Many such have been described in human beings. A few examples have been recorded in calves.

Up to this point we have been concerned with anterior and complete dichotomy: now posterior dichotomy of the trunk claims consideration. A simple case is furnished in the green lizard (*Lacerta viridis*) shown in fig. 69. The animal has a bifid tail. To show that the bifidity is not due to injury and subsequent reproduction of the tail and consequent budding, it may be mentioned that reproduced tails are never so perfectly conformed as in this specimen and do not contain vertebræ; further, monstrous lizards have been recorded in which the cleavage had involved the trunk so as to produce the bodies and limbs of two lizards but only one head. In one specimen described and figured by Tiedemann, a double-bodied monstrous green lizard has a double, but fused head.

Passing from tailed animals, such as lizards, and turning to frogs, it must be clear to every one that if dichotomy occur, it need only be slight to produce reduplication of the pelvis with the limbs and associated parts. This is a matter of interest because many specimens have been incorrectly interpreted on account of this fact not being properly appreciated. It is also clear that if dichotomy lead to reduplication of the pelvis, four pelvic limbs should result. In typical specimens such is the case: two of the limbs being functional, and two usually occupying a median position on the ventral aspect, much smaller in size.

FIG. 70.—*Rana esculenta*, with supernumerary hind limbs.

A specimen of this nature was described in 1837 by Dr. J. van Deen. It occurred in a frog (*Rana esculenta*), (fig. 70), and was associated with a bifurcated condition of the termination of the alimentary canal. Van Deen gives a carefully detailed account of the dissection of this frog. It seems to be one of the earliest recorded cases of this malformation. A much commoner condition is to find the abnormal limb projecting posteriorly between the normal legs. In such a position the two legs are as a rule fused together near their attachment to the trunk, but may retain their individuality in the distal segments. Such a specimen is sketched in fig. 71. It is a lamb with a supernumerary pair of confluent hind limbs, as shown by dissection, projecting from the

FIG. 71.—A Lamb with coalesced supernumerary pelvic limbs. (After Gurlt.)

pelvis. In this case the nether opening of the alimentary canal is double also.

A much more obvious case than this is represented in the chick (fig. 72). For some reason or other supernumerary legs are very common in fowls, ducks, geese, pheasants, and lambs.

In many cases posterior dichotomy of the axis may be indicated by one limb only, the other undergoing suppression.

Dr. Tuckerman [1] has placed on record an account of such a specimen. The subject was a frog (*Rana palustris*); it was blown from out a crevice in a ledge of mica schist during some blasting operations. The crevice was twelve feet below the surface and measured only a few lines at its widest point; flowing into the

FIG. 72.—A Chick with supernumerary legs.

crevice was a small stream of running water, which undoubtedly conveyed either the eggs, or the frogs in the larval stage, to the interior of the rock, as the breach in the ledge was much too small even to admit the passage of a very young frog. This frog had, as

[1] *Journal of Anatomy and Physiology*, vol. xx. p. 516.

represented in fig. 73, a third hind limb projecting posteriorly between the normal legs. On dissection it was found attached by muscle and tendon to the symphysis pubis: it presented the usual segments of femur, crus, and pes, and was furnished with two well-formed digits and the rudiment of a third. During life it was found that when the skin of the limb was stimulated the frog would jump quickly away; sometimes oft-repeated stimulation would call forth a muscular contraction in the limb. Occasionally movements were observed in the abnormal limb when the frog was left entirely to itself.

It has already been mentioned that in parasitic fœtuses, even when the limbs are well-formed, the autosite has rarely the power of spontaneously moving the limbs belonging to the parasite, and this is the rule in supernumerary limbs, so that the spontaneous movement in the leg of Tuckerman's frog is interesting. The drake (fig. 74) possessed the power of moving the abnormal limbs, and similar specimens have come under my notice in hens. Such power is, however, exceptional.

FIG. 73.—A Frog (*Rana palustris*), with supernumerary hind limb. (After Tuckerman).

Supernumerary pelvic limbs similar to those of frogs and toads, sheep and birds, occur in the human subject; several of the individuals thus affected have lived to twenty, thirty, or more years. Of the more remarkable cases may be mentioned Jean B. dos Santos of Portugal, and Blanche Dumas. The mode of origin of paired pelvic limbs has been especially studied by Professor Cleland. This anatomist has clearly demonstrated that many of these are due to posterior

FIG. 74.—A Drake with a pair of supernumerary legs.

dichotomy leading to reduplication of the pelvis. He further points out that in living specimens of posterior dichotomy, the functional limbs belong to opposite pelves. In some it will be right limb of the right pelvis and the left limb of the left pelvis. In such a case the adjacent sides of the pelves and the associated limbs become dwarfed or suppressed. In some the legs remain separate; in others they coalesce completely or partially, and in rarer specimens one leg may be suppressed. The relation of the suppressed to the functional pelvis varies greatly. In many specimens it is represented by an irregular plate of bone carrying a limb or limbs wedged in the pubic arch; in others two fairly-formed pelves may occur on the same plane. The relation borne by the opposed halves of these reduplicated pelves is similar to that which occurs in anterior dichotomy. For instance, we may have a single body with two perfect heads, or the dichotomy of the head is so partial that the jaws in apposition with each other, that is the left one of the right head and the right jaws of the left head, fuse together and form a composite pair of jaws wedged between two functional ones. In cases of partial posterior dichotomy, it has been already stated that the functional legs often belong to distinct pelves, so in many instances of anterior dichotomy the functional jaws belong to two heads. In anterior dichotomy, when it extends into the thoracic region, each individual has a pair of functional arms. In a few cases the median arms, that is the left arm of the right body and the right arm of the left body, fuse together and form a composite median limb.

Although apparently unconnected with the subject of twin monsters, it is necessary to offer a few remarks on the subject of transposed viscera. Very many cases have now been observed and dissected in which the heart, instead of being inclined to the left side is turned to the right, the liver is situated in the left side of the abdomen, thus changing places with the spleen. This transposition also affects the main arterial and venous trunks. Mr. Morrant Baker suggested to me that if double embryos and, in some instances, twins are produced by dichotomy of a single ovum, the viscera of one or other embryo should be transposed—one should be as it were the reflection of the other *velut in speculo.*

There is much to support this view. In many examples of duplex human children such a transposition has been actually found; in following up the inquiry it turns out that the hearts in these monsters are often situated in the median plane of the body, and not infrequently a composite heart dominates the circulation of the two individuals. The question naturally suggests itself, Are all cases of transposition of viscera associated with twin conceptions? This must stand as a question until we know more of the history of such cases, and it needs careful investigation.

It was formerly believed, and some still maintain the view, that duplex monsters may arise by the adhesion of two embryos originally distinct. This view has little foundation. Without attempting to seriously discuss the difficulties surrounding this "impaction theory," its supporters will have to explain the relation of such fœtuses to each other, and why they are united by homologous parts, such as head to head, back to back, pelvis to pelvis, &c. Should the second embryo be represented merely as a teratoma, it usually resembles the part to which it is attached. By the impaction theory it is inconceivable that these parts should accurately adapt themselves to each other, that heads should unite, hearts blend, or the alimentary canals fuse to form one. On the dichotomy or cleavage theory all this becomes intelligible, and the fact that reduplicated digits, limbs, heads, and trunk occur throughout the vertebrate sub-kingdom and beyond it, is sufficient to indicate that a common cause underlies these diverse malformations.

The subject is one of great interest, for the same tendency which produces dichotomy of the ray in starfish, or digits in mammals, will, when it involves the axis of the limb, produce a supernumerary arm, wing, or leg; should it affect the axis of the embryo, will lead to the production of duplex monsters of varying development and different degrees of union, or even result in viable twins. The same tendency to dichotomize is exhibited in the plant world.

CHAPTER VI.

ATAVISM OR REVERSION.

MUCH that is fanciful and speculative is mixed up with the subject of atavism; the widespread acceptance of the principles of evolution has had the effect of rendering us less critical in the examination of suspected cases. On the present occasion atavism or reversion will be considered only in relation with structural peculiarities, and an endeavour will be made to indicate spurious examples of the process.

Gegenbaur defines atavism as a "reappearance of a more primitive organization, or a reversion to a primary state." He is also careful to point out that atavism does not consist in the existence of a latent germ, but in its becoming perfected or further developed. This view may be more clearly expressed thus:—

Atavism consists in the attainment of a functional or, more or less, full development of parts which for a given animal are suppressed during embryonic life, or undergo great modification.

This definition allows atavistic phenomena to be arranged in two groups:—

1. *The attainment of a functional condition by structures normally suppressed.*
2. *Reversion of organs and tissues to an original type.*

Examples of each form will now be considered. The attainment of a functional condition by parts normally suppressed is well illustrated in the case of man by supernumerary ribs. Normally the number of ribs on each side is twelve: not infrequently the number is increased by an additional rib at the cervical or lumbar end of the series. In many birds and reptiles all the cervical vertebræ bear ribs; in mammals the cervical ribs are represented in a vestigial form, and in man such vestiges are confined to the fifth, sixth, and seventh vertebræ. Occasionally the rib attached to the seventh cervical vertebra acquires a length of five to seven centimetres, becoming perceptible, to practised eyes, even in the living body. Supernumerary ribs attached to the lumbar vertebræ are more instructive than those in the neck. Each normal rib has inserted into its angle a fan-shaped muscle which raises it during inspiration and is known as the *levator costæ*. In the lumbar region these muscles are represented by fan-shaped collections of tendons connected with the transverse processes of the vertebræ. In many instances, when a lumbar rib is present, the corresponding tendinous levator is replaced by a functional muscle.

This is instructive, as it illustrates a fact which must be constantly borne in mind when investigating suspected cases of reversion, viz., *atavistic parts do not belong to forms palæontologically remote or systematically far distant.* This is particularly insisted upon by Gegenbaur.

In the present case the lumbar ribs are represented by extra centres of ossification at the extremities of the transverse processes; normally they fuse with these processes, but occasionally attain a functional size and are recognized as supernumerary ribs. In order to find mammals with more than twelve pairs of ribs we need not seek far from man, as the chimpanzee and the gorilla possess thirteen pairs; in the American monkeys there are twelve to fifteen pairs, and in lemurs twelve to sixteen pairs.

The following specimen illustrates the reversion of

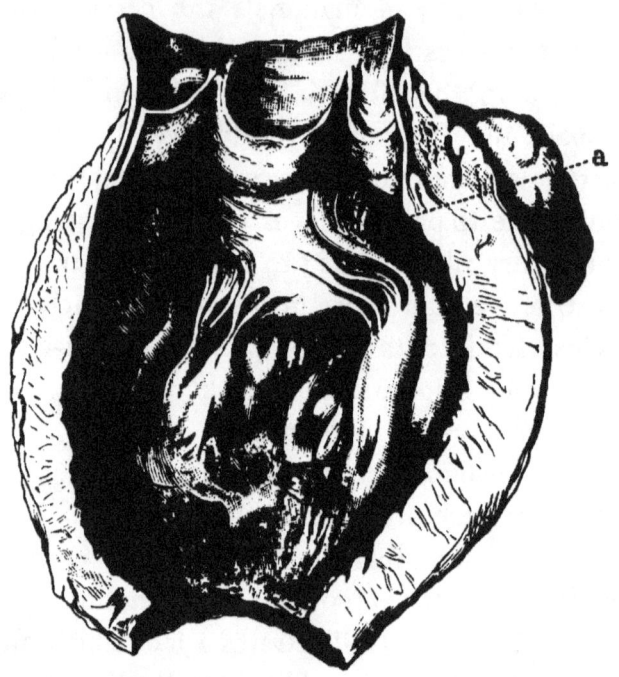

Fig. 75.—The mitral valve of a man's heart containing a patch of muscle tissue, *a*. (After Dr. Ogle.)

tissue. Dr. Ogle [1] has described a man's heart, which contained a patch of muscle tissue in the anterior segment of the mitral valve (fig. 75). The patch of muscle was as large as a fourpenny-piece, and resembled under the microscope the natural tissue of the heart.

[1] "Trans. Path. Soc.," vol. ix.

In the embryo these flaps, or valves, arise as outgrowths from the cardiac walls, and consist of muscle tissue, which gradually undergoes metamorphosis into fibrous tissue. In the echidna and ornithorhynchus the valve of the right ventricle is muscular throughout life : this is also the case in all birds. Hence a patch of muscle tissue in the cardiac valves may be regarded as atavistic.

The arch of the aorta, the main arterial trunk, now and then illustrates this form of atavism. In all mammals the aorta in its normal course should curve over the root of the left lung; in birds, with equal constancy, it curves over the root of the right lung; in lizards and frogs an aortic arch crosses the root of each lung and, subsequently uniting, form a single trunk. In the embryo of birds and mammals both aortic arches exist, but in mammals the right one becomes suppressed ; occasionally, however, the left arch, from causes quite unknown to us, undergoes suppression, the right one becoming the functional vessel. Thus a right aortic arch in a mammal is a true instance of atavism.

It is well known that in the small amount of hair upon the body, man differs from all mammals except the whales and porpoises. He however stands alone in its unusual distribution. The human fœtus is covered with delicate hair known as lanugo, which is usually shed shortly before or immediately after birth. This is clearly an indication of a lost character, the lanugo representing the hairy covering of closely allied vertebrates ; it appears in the embryo in obedience to heredity. Children are often born with pigmented hairy patches on their bodies known as moles ; these moles vary in size; sometimes they are no larger than split peas, in other cases cover several square inches, and in exceptional cases nearly the whole of the trunk may be thus covered; the hair may even equal in length that growing on the child's head. Several instances of hairy men of this sort have been known, and an illustrative case is shown in fig. 76. The original drawing is preserved in

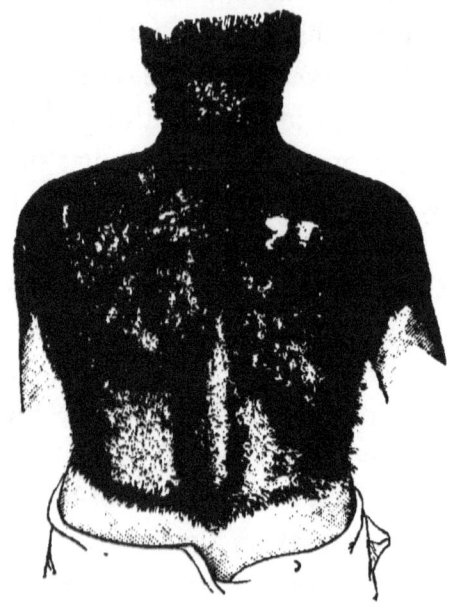

FIG. 76.—Trunk of a Man with a large hairy mole. (After Lawson.)

the museum of Middlesex Hospital. The biblical Esau was probably a man of this character. These are examples of reversion, the lanugo instead of being shed attains full development: we need only descend to those mammals most closely allied to man, and in his own order Primates, to find the true significance of the phenomenon.

Some instances of the reversions of organs and tissues to an original type will now be considered, and it will serve to simplify matters if the question be illustrated by a botanical example of this form of atavism. In 1790 Goethe, in his celebrated treatise " On the Metamorphosis of Plants," drew the attention of botanists to the fact that the various parts of a flower may be regarded as modified leaves. Of course this does not mean that each part of the flower is a metamorphosed leaf, but that we are able to trace every structural gradation from leaves to bracts, from bracts to sepals, and not infrequently sepals will be replaced by, or become converted into, true leaves. The changes from sepals to petals and from petals to stamens are as gradual as from bracts to sepals, and the homology is often declared by the stamens becoming petals, and petals appearing as leaves, even stamens have been known to revert in this way; this is what is meant when we say that the various parts of a flower are formed upon a common type.

Atavistic phenomena of this character are not unknown in the animal world, some of which we will now consider.

One of the most remarkable recorded examples of this form of atavism occurred in a rock-lobster (*Palinurus penicillatus*). It is the specimen originally described by M. Alphonse Milne-Edwards, and referred to by Darwin; the drawing has been published by Professor G. B. Howes (fig. 77).

Notwithstanding the differences observed in the various appendages of lobsters, such as the antennæ, eye-stalks, swimmerets, chelæ, and the like, morphologists hold the opinion that all these appendages are modifications of a common type. Normally the eye-stalk corresponds to the protopodite of an abdominal limb, consisting of a short basal and a long cylindrical terminal joint, the distal surface of which is covered with corneal facets. In A. Milne-Edward's *Palinurus* we find a normal ophthalmite on the right side, but the left one "has taken on antenniform characters."

FIG. 77.—Cephalon of a Rock-lobster (*Palinurus penicillatus*), with an antenniform process growing from the inner aspect of the eye-stalk. (After Howes.[1])

Professor Howes points out that the corneal facets on the eye-stalk of decapod crustaceans do not in many instances surmount the whole of the free surface; frequently the outer free border is destitute of corneal facets, and often is swollen and well differentiated. This is so in *Palinurus*, and would serve to support the view that the facetted portion of the oph-

[1] "Proc. Zool. Soc." 1889.

thalmite was exopoditic, whilst the inner portion, which furnished the antenniform process, was endopoditic in character.

The interest of this specimen lies in the fact that it is exactly analogous to the conversion of an element of a flower, say a stamen, into a leaf. Although I have made a wide search, nothing in any way comparable to this specimen can be found recorded in zoological literature, but there can be little doubt that such cases occur if we would watch for them. One would also imagine that similar malformation must occasionally occur in the appendages of insects, but thus far my inquiries have not had any material result.

From organs we turn our attention to tissues, and examine some specimens illustrating the reversion of mucous membrane to skin.

The exterior of our bodies is covered by a membrane called skin, which differs in many points from that lining the internal cavities—the mouth, stomach, and intestines—and known as mucous membrane. The contrast between skin and mucous membrane is well illustrated in the case of the conjunctiva, the delicate, sensitive membrane covering the eyeball and ocular aspect of the eyelids. The conjunctiva is really modified skin, and not infrequently declares its ancestry by reverting to its original form. Indeed, it is no unusual event to find a patch of hair-covered skin growing upon the ocular conjunctiva. Such cases are far from rare. Mr. T. Collins is of opinion that about twelve cases of this abnormality are seen annually at the Ophthalmic Hospital, Moorfields. These dermoid patches are most frequent on the outer side of the eye near the junction of the cornea and sclerotic.

Similar hairy patches have been recorded as occurring on the conjunctivæ of dogs, oxen, and sheep; in the latter animals these abnormal pieces of skin are furnished with wool. The specimen sketched in fig. 78 represents a large patch of skin growing on the con-

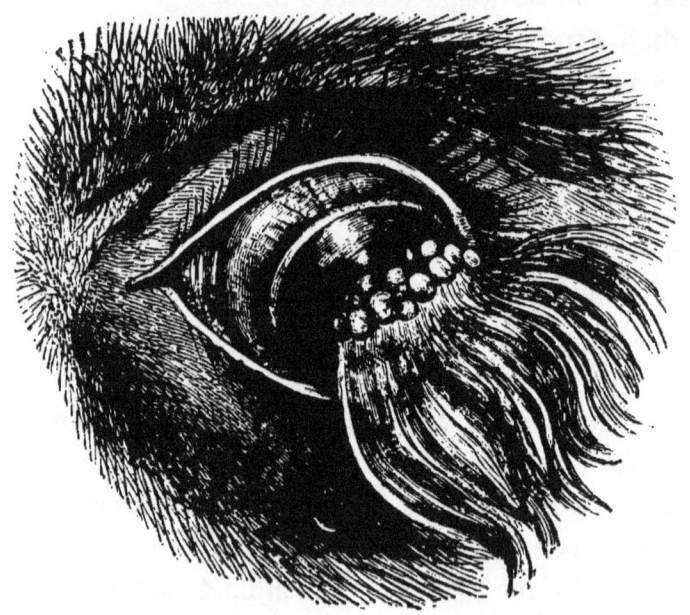

FIG. 78.—The cornea of an Ox with a patch of piliferous skin growing from it. (After Partridge.)

junctiva of an ox. Some of these patches of skin have been examined microscopically and found to contain parts which are characteristic of skin in other situations of the body such as pigment, sebaceous, and sweat-glands. The developmental details of the conjunctiva indicate most conclusively that it is a modified piece of skin, and though in the individual

ATAVISM OR REVERSION.

it does not pass from skin to mucous membrane, nevertheless, as in the case of the flower, it now and then indicates its ancestry by reverting to its original type of structure.

Spurious Atavism.—It is so important not to con-

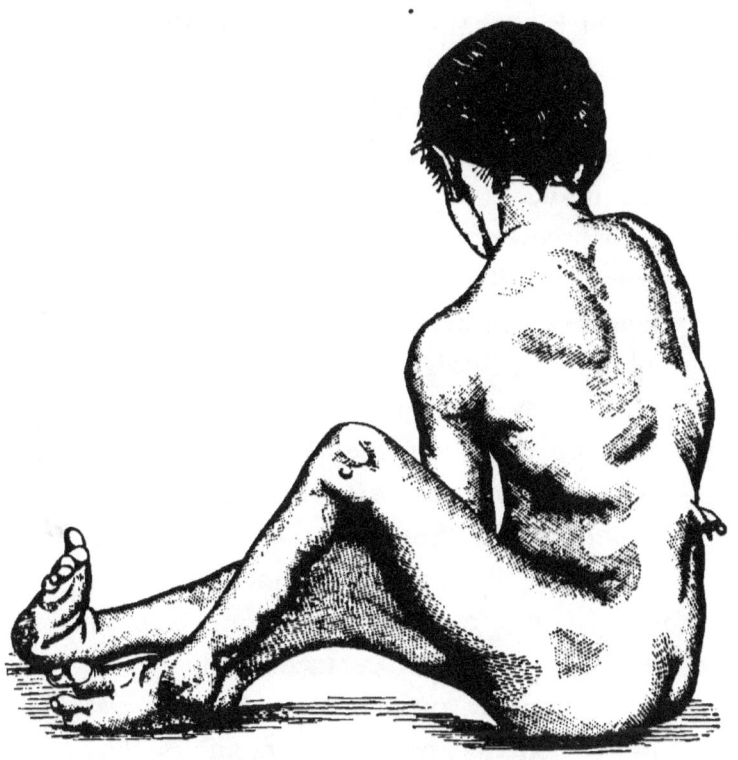

FIG. 79.—A congenital fold of skin stretching from the thigh to the heel of a girl. (After Wolff.)

found spurious instances with examples of genuine reversion or atavism that, even at the risk of being tedious, it is necessary to consider briefly a few interesting cases likely to be misinterpreted on superficial examination.

In 1888 Dr. Julius Wolff, of Berlin, published a

description of a remarkable anomaly in the leg of a girl. Among other defects she had a broad triangular fold of skin stretching from the thigh to the heel, as shown in fig. 79. The malformation was congenital and is in all probability unique; at any rate, it is very rare, for such exhaustive writers on the subject of malformations in general as Förster, St. Hilaire, Alhfeld, and Albers furnish no parallel case.

On superficial examination a zealous evolutionist might be induced to argue that we have here an attempt to produce a wing, or regard it as a reversion to the parachute-like folds of skin seen in phalangers, flying squirrels, or even the wings of birds. Wolff, however, judiciously disposes of fanciful speculation in this direction by correctly pointing out that, in the so-called flying mammals, the cutaneous folds, or parachute, extends from the fore to the hind limbs, and this was the condition in those extraordinary extinct forms the pterodactyls; indeed, no mammals are known with skin-folds passing from the two segments of the hind limb. In birds we find a cutaneous fold passing from the humerus to the carpus and known as the patagium, which in its general appearance resembles the fold in the girl's leg. Beyond this superficial resemblance these patagii have nothing in common, for the wing-fold in the bird is traversed by tendons some of which are remarkable for their elasticity. Thus we cannot find among mammals, birds, or lizards, living or extinct, anything corresponding to this curious wing-like expansion. We must therefore regard this skin-fold as a spontaneous variation, or "sport." Even if an animal were known

ATAVISM OR REVERSION.

to possess such a web, there would be another strong argument against its reversionary character, viz., that in the embryo the two segments of the limb are not united or in any way connected by a cutaneous fold; thus we have not to deal with a persistent fœtal structure.

Whilst dealing with skin-folds it may be of interest to draw attention to a not infrequent example of webbing. In the human subject it is not uncommon to meet with individuals possessing two or more fingers united to an abnormal degree by a web. Normally our fingers are webbed as low as the middle of the first phalanx. The gorilla has a web extending to the first joint of the index and middle finger, and the Siamang gibbon has its second and third toes webbed as far as the distal end of the first phalanx; the seal and ornithorhynchus have a broad web to the digits of manus and pes.

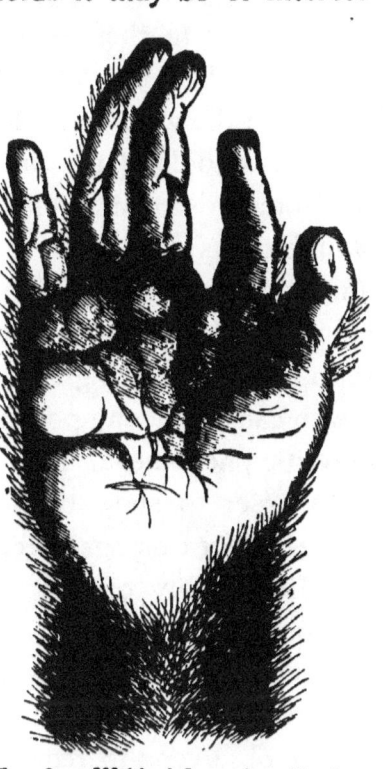

FIG. 80.—Webbed finger in a Monkey (*Pithecus satanus*).

Abnormally webbed fingers and toes are not peculiar to man; they occur in monkeys, and a good specimen of this malformation which occurred in a monkey (*Pithecus satanus*) is drawn in fig. 80. In this instance the corresponding fingers on each hand were similarly affected. In such cases as this it is easy to argue that such conditions are reversions to the webbed feet and hands of aquatic mammals, but careful consideration leads me to take an opposite view. If all the fingers and toes were furnished with a web we should be justified in regarding the phenomenon as atavistic, for in their early stages our fingers are completely webbed, but later the digits grow at a greater rate than the web, until the skin-fold reaches no lower than the middle of the first phalanx. This would lead me to regard the abnormal union of fingers and toes by skin-folds as a sport or spontaneous variation, and not atavistic; and this opinion is strengthened by the circumstance that, as we have just seen, a patagium may form between the thigh and leg, a situation where such a web cannot by any possible means be classed among atavistic phenomena.

In order to show to what absurd conclusions loose modes of reasonings lead, reference may be made to the abnormality of the stomach known as congenital or hour-glass contraction. In such cases the human stomach is divided by a constricted or narrow portion into two compartments (fig. 81), and this has induced a few observers to regard this as a reversion to the complex stomach of the ruminant, especially the two-chambered stomach characteristic of the genus *Cervulus*.

There is, however, the following difficulty in accepting this explanation. All such deformed stomachs have been observed in adults; no one has ever found a double-chambered stomach in a child. Again, in many of the specimens evidence of disease, such as ulcers, old and recent, or scars have been detected in the

ATAVISM OR REVERSION. 147

neighbourhood of the contracted portion. During its development the human stomach never presents itself otherwise than as a single-chambered viscus. This is in itself an important fact, for if in the embryo the stomach presented a median contracted portion it could easily be conceived that in some cases the contraction would persist. It must also be borne in mind that the ruminant's stomach is highly specialized; there-

FIG. 81.—A Human Stomach with hour-glass contraction.

fore, as reversion consists in the reappearance of a lost character we shall have to show that mammals closely allied to man possess such a stomach : this is not the case. Thus embryology and comparative anatomy are eloquent in denying the reversionary nature of hour-glass contraction of the human stomach.

Having attempted to indicate the general grounds upon which suspected atavistic phenomena, in so far as malformations are concerned, should be analyzed, it will perhaps be interesting to discuss reversionary phenomena as they manifest themselves in relation with a particular organ or group of organs; it will in this way be possible to display, in a striking manner, the large amount of misconception which prevails in regard to atavism in general. This will serve to show that reversion of organs or tissues, and the assumption of function by suppressed parts, is not so frequent as many suppose, especially when each suspected case is submitted to rigid analysis. I shall, however, first of all consider atavism in relation to secondary sexual characters, and then proceed to the study of supernumerary digits and limbs, and supernumerary mammæ and nipples. It will be to the reader's advantage before perusing those sections to study carefully the remarks made on dichotomy, for of all examples mistaken for atavism, dichotomy claims by far the greatest number.

Atavism in relation to Secondary Sexual characters.— As Darwin points out, two distinct elements are included under the term "inheritance"—the transmission and the development of characters. The distinction is so important, especially in its bearing on the question of atavism, that the two conditions will be illustrated by concrete examples.

In most species of the deer tribe it is the rule for the male alone to possess antlers, yet it is a well attested circumstance that under certain diseased conditions of the sexual organs, especially atrophy or degeneration of the ovaries, rudimentary horns which are never shed appear in the female.

This shows us that the female is in possession of secondary sexual organs in virtue of transmission, yet they remain latent as a rule, and only become developed under extraordinary circumstances. The same holds good for those cases of hens who for years lay eggs, yet eventually cease to do so, put on one side the plumage proper to their sex, and adopt more or less completely that of the cock.

Hunter[1] recorded some examples of this; the following is one of them :—

"Lady Tynte had a favourite pyed pea-hen which had produced chickens eight several times. Having moulted when about eleven years old, the lady and family were astonished by her displaying the feathers peculiar to the other sex, and appeared like a pyed peacock. In this process the tail, which became like that of a cock, first made its appearance after moulting; and in the following year, having moulted again, produced similar feathers. In the third year she did the same; and, in addition, had spurs resembling those of a cock. She never bred after this change in her plumage, and died in the following winter during the hard frost, in the year 1775-6."

The change here described, and now seen in a large number of birds—indeed, I know of no gamekeeper of experience who has not seen such specimens—has been shown conclusively to be associated with atrophy or non-development of the ovary. The best marked case that has come under my observation occurred in a hen golden pheasant. I watched the bird for two years. She presented the resplendent dress of the cock, but her plumage was not quite so brilliant, had no spurs, and the iris was not encircled by the ring of white so conspicuous in the male. Coincident with the change of plumage she ceased to lay eggs. After this transformation she lived happily with her mate for two years, and in spite

[1] An Account of an Extraordinary Hen Pheasant ("Observation on the Animal Œconomy").

of the deceit of feathers I have seen a pheasant in the next cage making her overtures. Two years later (the spring of 1885) she would not yield to the solicitations of her mate, and he savagely killed her. The ovary was of the size of a split pea. A drawing of this bird, with the assumed plumage, is given as a frontispiece, and the hen golden pheasant in the proper sober plumage is sketched for comparison.

These specimens are of interest in many ways, but they are especially instructive in showing how limited our knowledge is concerning latent characters; although this comes out in a clear manner in the case of the golden pheasant, it is even more forcibly illustrated in an example mentioned by Darwin. It is to this effect :— " Mr. Hewitt possessed an excellent Sebright gold-laced hen bantam which, as she grew old, became diseased in her ovaria, and assumed male characters. In this breed the males resemble the females in all respects except in their combs, wattles, spurs, and instincts, hence it might have been expected that the diseased hen would have assumed only those masculine characters which are proper to the breed, but she acquired in addition well arched tail, sickle feathers quite a foot long, saddle feathers on the loins, and hackles on the neck—ornaments which," as Mr. Hewitt remarks, " would be held abominable to her breed." The Sebright bantam is known to have originated about the year 1800 from a cross between a common bantam and a Polish fowl, re-crossed by a hen-tailed bantam, and carefully selected; hence there can hardly be any doubt that the sickle feathers and hackles which appeared in the old hen were derived from the Polish fowl or common bantam, and there we see that not only certain masculine characters proper to the Sebright bantam, but other masculine characters derived from the first progenitors of the breed removed by a period of above sixty years were lying latent in this hen-bird ready to be evolved as soon as her ovaria became diseased." (Animals and Plants under Domestication.)

These examples open up the subject of secondary sexual characters. The question of primitive hermaphroditism has been already discussed in a preceding chapter, and an attempt was made to show that, for a brief period at least, the embryo presents sexual parts common to the male and female, so that for a time it is absolutely impossible to determine the sex. What is true of the embryo applies equally to animals normally hermaphrodite: no distinctive characters are displayed externally. Also in cases of hermaphroditism occurring in animals normally bisexual, the secondary sexual characters are intermediate to those of the functional male and female. It is therefore fairly evident that the female, though she differs from the male in the non-development of secondary sexual characters, yet possesses them in a latent condition; or, to put the matter briefly, they are transmitted, but not developed.

We must now inquire how it is, that if the female possesses all the secondary sexual characters of the male in a latent manner, they are prevented manifesting themselves.

When differentiation of sex occurs in animals previously hermaphrodite, it involves either the loss of certain characters on the part of the female, or the acquisition of new characters by the male, or at any rate increased functional importance of certain organs possessed, when in the state of hermaphroditism, by all the forms. By natural selection the male would acquire (or, if already in his possession in a functional condition, would gradually develop) means for seizing and retaining the female, such as the claspers of sharks, the callous pads of frogs, &c. Paternal duty requires the male to protect the young and defend the females from harm; hence horns, teeth (as in the musk-ox), spurs, tusks, &c., become more developed in him.

The duties of the female require her not only to furnish the material out of which the young are to be formed, but in many cases she is required to provide them with nutrition long after they enter the world. The material which the female thus provides is of the very kind necessary, in many instances, to build up such structures as horns, tusks, teeth, and the like. Further, this material is required by the female at the corresponding period of life in which these structures become developed in the male, viz., on the advent of puberty. We may state with certainty, that a distinct correlation exists between the generative organs of the female and the development of the secondary sexual male characters. The more developed and functional the female reproductive organs become, the less likely is she to manifest the secondary characters of the male. It may be argued, that in some cases the female simulates the male, as in the few examples of female deer possessing horns. Quite true; but so long as the female is engaged in the duties of reproduction, these secondary characters are never developed to the same extent as in the functional male. It must also be borne in mind, that in cases where sterile females, or those which have ceased to bear young, put on external male characters, they rarely attain such proportions or beauty as in the male ; for in the males the general excitement produced by sexual passion has a most powerful stimulant effect upon the growth and development of these structures, which is wanting in the female. So that in her occasional attempts to emulate the male she succeeds to a certain degree, but rarely, if ever, attains to so good a condition.

Hunter has recorded some experiments which have a bearing on this matter :—

" I wished also to ascertain if the parts peculiar to the male could grow on the female, and if the parts of a female, on the contrary, would grow on a male.

Although I had formerly transplanted the testicles of a cock into the abdomen of a hen, and they had sometimes taken root there, but not frequently, and then had never come to perfection, yet the experiment could not, from this cause, answer fully the intended purpose ; there is, I believe, a natural reason to believe it could not, and the experiment was therefore disregarded. I took the spur from the leg of a young cock, and placed it in the situation of the spur in the leg of a hen-chicken ; it took root, the chicken grew to a hen, but at first no spur grew, while the spur that was left on the other leg of the cock grew as usual. This experiment I have repeated several times in the same manner, with the same effects, which led me to conceive that the spur of a cock would not grow upon a hen, and that they were, therefore, to be considered as distinct animals, having very distinct powers. In order to ascertain this, I took the spurs of hen-chickens and placed them on the legs of young cocks. I found that those which took root grew nearly as fast, and to as large a size as the natural spur on the other leg, which appeared to be a contradiction to my other experiments. Upon another examination of my hens, however, I found that the spurs had grown considerably, although they had taken several years to do it; for I found that the same quantity of growth in the spur of a cock, while on the cock, during one year, was as much as that of the cock's spur on the hen in the course of three or four years, or as three or four to one; whereas the growth of the hen's spur on the cock was to that of the proper spur of the hen as two to one."

When a female animal belonging to a dimorphic species assumes male characters, it is truly an example of atavism, or development of transmitted characters normally latent.

Before leaving this matter it will be useful to indicate to some extent our knowledge as to the frequency of the assumption of male characters by the females of dimorphic forms. In birds there can be no question that very many specimens have been seen and carefully studied, but in the case of female deer putting up antlers the case is very different. On making inquiries of persons likely to have seen such specimens I find it extremely difficult to obtain authentic information, and many of the statements in circulation on this matter have very shallow foundations. One of the most accessible specimens is preserved in the museum of the Royal

ATAVISM OR REVERSION.

College of Surgeons, London. It is the skull of a doe roedeer; it was shot by the Earl of Egremont, near Petworth, Sussex, in 1810, and presented by him to the museum. The antlers, as shown in fig. 82, were evidently covered with "velvet." This specimen is described by Mr. E. R. Alston.[1] One is a simple curved snag, nearly eight centimetres in length, with a well-developed burr; the other is a mushroom-shaped burr without any beam. Lord Egremont, in

FIG. 82.—A female Roedeer (*Capreolus capræa*), with antlers. (After Alston.)

a letter, expressly stated that the deer was "a very old and uncommonly large female with two young ones in her." Alston states that in Germany, where the roedeer is more plentiful than in this country, many does with antlers have been recorded, no fewer than forty instances being known to Dr. Altum. Most of these were barren animals, and the antlers were always of a more or less abortive character, except one

[1] "Proc. Zool. Soc." 1879.

case in which the normal male form was well reproduced; but several were fertile and either with young when killed or had recently given birth to fawns. Such abnormal antlers appear to be always persistent and permanently covered with velvet.

Outside the genus *Capreolus* (excepting, of course, the genus *Rangifer*, the females of which are normally fur-

FIG. 83.—Head of a female Moose (*Alces machlis*), with antlers.

nished with antlers) it is rare to find antlered females. In fig. 83 a drawing of the head of a female moose (*Alces machlis*) is given with velvet-covered horns; for the details of this specimen I am indebted to Mr. H. E. Dresser:—
"The female moose was shot by an Indian, on the Upper Musquash, and taken into Indian Town, St. John's, N.B., where it was sold to a butcher; I was then in charge of

the Musquash estate, and hearing that a cow moose with horns had been shot I at once drove over to St. John's, saw the beast which was then unskinned and of course I could not see that it was a cow, so I bought the head for a trifle, skinned and stripped it. This was in November, 1859."

We have authentic evidence of the occurrence of antlers in female deer—in roedeer (*Capreolus capræa*), Virginian deer (*Cariacus Virginianum*), moose (*Alces machlis*), red deer (*Cervus elephas*). The occurrence of antlered females in other genera of deer is extremely rare.

The existence of antlers in female deer is not, as far as I can ascertain, necessarily associated with ovarian disease or sterility. Again, hen birds which assume cock plumage regularly moult; female deer with antlers do not shed them as is the case with the males which they mimic.

It may be useful to add a list of birds in which the hen has been seen in male plumage: pheasants (common, golden, and silver), the common hen, pea-hen, partridge, bustard, American pelican, wild and domestic duck, cuckoo, cotinga or bell-bird, chaffinch, bunting.

CHAPTER VII.

ATAVISM (*continued*).

Supernumerary Digits, Limbs, and Mammary Glands.—Nothing illustrates so forcibly the necessity of critically examining suspected cases of atavism as those abnormalities collectively known as Polydactyly. At the outset it may be stated, contrary to the prevailing opinion, that supernumerary digits are very rarely atavistic.

The various examples of extra fingers, toes, arms, and legs, described in the chapter on Dichotomy, serve to show that if all supernumerary parts are to be regarded as reversions we must find vertebrates provided with an unlimited number of toes, double hands and feet, and more than two pairs of limbs.

No vertebrate animals other than fish and the *Ichthyosaurii*, possess on each limb more than five digits, therefore when the number of toes or fingers exceed on each limb this typical number, it must if we, with Darwin, regard the accessory digit as atavistic, be a reversion to an Ichthosaurian or a fish form. The distance is far too great, and in doing so we violate the rule that atavistic parts do not belong to forms palæontologically remote or systematically far distant.

ATAVISM.

Among reptiles, birds, and mammals, our knowledge at present only warrants us in regarding extra digits as reversions when they occur in non-pentadactyle animals. For instance, the spider monkeys have no thumbs; a careful dissection of the hand, however, will reveal, in connection with the trapezium, a band of fibrous tissue containing nodules of hyaline cartilage, representing the missing thumb. On one occasion I was surprised to find in an adult specimen of *Ateles panniscus* a small thumb projecting above the web of the finger and furnished with a nail. This condition is, I understand from competent zoologists, not infrequent, and as it is an example of the attainment of a functional condition by an organ usually suppressed in this species, it comes within the definition of atavism. Further, those monkeys, its closest allies, marmosets, *Mycetes*, and *Callithrix*, possess functional thumbs.

All cases of extra digits in non-pentadactyle mammals are not necessarily atavistic: let us consider this in reference to the horse. That the modern horse walks upon the enlarged third digit, and has a vestigial metacarpal or metatarsal on each side of it, is accepted by morphologists. The comparatively recent ancestors of the horse had three functional digits. Hensel's investigations on *Hipparion mediterraneum* indicate the probability that the inner (second) digit was the last to abort. Horses are occasionally seen with two functional digits instead of one, and it is a noteworthy circumstance that in the majority of cases it is the inner digit which reappears, that is, the one which we should theoretically expect to reappear most frequently. Such cases are truly atavistic, and the leading features in the argument are visually represented in the drawings (fig. 84).

The museum of the Ecole Vétérinaire de Lyons contains a very remarkable specimen. It is the skeleton of an equine manus with two functional digits (fig. 85), on each side of the metacarpal bone a vestigial metacarpal exists as in the normal manus. In the specimen last considered we noted that the supernumerary digit was articulated with the distal end of the lateral metacarpal (fig. 85, B). In the Lyons speci-

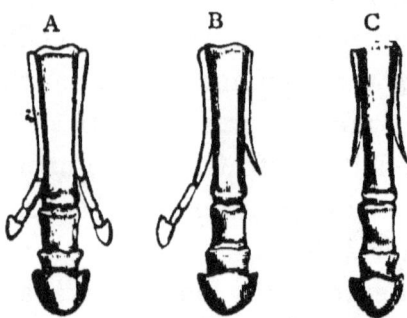

FIG. 84.—A, the manus of *Hipparion*; B, the manus of a horse (*Equus*) with the inner digit functional (atavistic); C, the normal manus of *Equus*, one toe functional, ii. and iv. vestigial.

men both digits articulate with the main metacarpal; and it is clear from even a cursory examination of the preparation that we have to deal with an example of dichotomy, or bifurcation of the terminal segments of the functional digit.

The museum of the Royal College of Surgeons, England, contains a skeleton of an equine manus with a supernumerary digit of this character, but it differs from the Lyons specimen in that the third digit has dich-

ATAVISM.

otomized unequally, and is thus more likely to be erroneously interpreted from a superficial examination.

Bardeleben has recently attempted most strenuously to prove that the ancestors of modern mammals were heptadactyle rather than pentadactyle, and he bases this opinion on a supposed discovery in the human embryo of some remnants of the missing digits, on the pre- and post-axial side of the foot respectively. In addition he draws attention to the existence in some mammals, chiefly rodents, of an ossicle in the foot, usually regarded as a sesamoid. This Bardeleben regards as vestiges of the missing digits on the inner side of the manus and pes, and terms prepollex and prehallux respectively. On these grounds he would urge that supernumerary big-toes and thumbs are atavistic.

Even supposing the vestiges of such digits to exist as Bardeleben believes, it would influence but little the views now held regarding the nature of bifid thumbs and big toes, for by an overwhelming amount of evidence it can be shown that in man they arise by dichotomy of the digits.

FIG. 85.—Supernumerary digit in the manus of a horse due to dichotomy. (Modified from Cheaveau.)

A full study of the question of supernumerary digits in all its bearings, and the examination of a very large number of specimens convinces me that an excess in the number of digits, in pentadactyle mammals, is due to dichotomy, and can in no sense be regarded as reversion. Supernumerary digits in animals with fewer digits than five are, in some few cases, due to atavism, but in many instances dichotomy is responsible for the abnormality, and it is only after careful dissection by an experienced anatomist that it is possible to determine to which of these causes the accessory digit or digits can be ascribed.

Supernumerary Mammæ and Nipples.—It is so difficult to distinguish atavistic phenomena from counterfeit manifestations that the great need for care in this direction will be further illustrated by discussing supernumerary mammæ and nipples. Up to the present time accessory mammæ, nipples, or teats, have been reported on reliable authority in the following situations:—the anterior wall of the chest, abdomen, axilla, arm, shoulder, the cheek, buttock, and even in ovarian dermoid cysts. If we accept unreservedly, that supernumerary mammæ are always atavistic, we force ourselves into the absurd position of imagining ancestors better supplied with milk-glands than even Diana of Ephesus; or if we protect our opinion by stating it in this way:— When a mamma occurs in an abnormal situation in man, but corresponds to the normal situation of the mamma in some lower animal of our class, it is then atavistic, we save ourselves much trouble.

In a few cases the guess would be correct, but in most it would be erroneous, and such loose methods of determining the nature of abnormal organs will in the long run lead to much confusion. At the outset we are faced by the question, What is a mamma? A mamma is a collection of modified sebaceous glands secreting milk, and having close relation to the function of reproduction. As a rule such glands are furnished with a cutaneous papilla, known as the nipple, or teat, which is either traversed by the excretory ducts of the gland, or forms a conduit whereby the young are enabled readily to suck the secretion. Mammary glands without ducts occur normally only in *ornithorhynchus* and *echidna*. Milk-glands are a distinguishing feature of mammalia, being unknown below this class.

Mammary glands are, as a rule, regularly and symmetrically arranged along the ventral aspect of the trunk, in two rows. When numerous they extend along the thorax and abdomen into the inguinal region. The teats usually correspond to the maximum number of young at a birth, but to this rule there are many exceptions. The greatest number known is fourteen pairs. When the number of functional glands is reduced to two, traces of the suppressed mammæ occur as supernumerary or accessory nipples, with or without rudimentary glands.

The reduction in number may take place in the thoracic region, leaving the inguinal set functional as in the cow, mare, sheep, goat, deer, and antelope. The inguinal glands may abort, leaving the pectoral set functional as in man, dugong, sloth, manatee, and monkeys. In marsupials the milk-glands are inguinal in position, and protected by the remarkable fold of skin which forms the pouch or marsupium.

The lemurs offer instructive modifications. The ring-tailed lemur (*L. catta*), the black lemur (*L. macaco*), and the mongoose lemur (*L. mongoz*), possess one pair of pectoral mammæ, as in man and quadrumana. Coquerel's mouse lemur (*Chirogaleus coquereli*), and the dwarf lemur (*Microcebus smithi*), possess three pair of mammary glands: one pair in the pectoral, another in the thoracic, and the third pair in the inguinal region (fig. 86). It is said, but I have had no opportunity of verifying the statement, that *Tarsius spectrum* has an inguinal and a pectoral pair of milk-glands; the aye-aye (*Cheiromys*) has only an inguinal pair; lastly, the

FIG. 86.—A Dwarf Lemur (*Microcebus smithi*), showing the arrangement of the mammary glands.

gray, or broad-nosed lemur (*Hapalemur griseus*), has the mammary glands situated on its arms.[1]

The above facts (with the exception of those relating to *Hapalemur*) seem to indicate that the lemurs have descended from ancestors possessing two rows of mammæ. In the dwarf lemurs, the abdominal pair

[1] Beddard, "Proc. Zool. Soc."

have disappeared, whilst in the ring-tailed, black, and mongoose lemurs the pectoral pair alone persist.

The persistence of the pectoral pair in these lemurs may possibly be explained by the curious manner in which they carry the young. *Lemur macaco* has usually one at a birth, and this is carried about entwined around its mother in the manner represented in fig. 88. I have watched the young lemur, and find that it rarely leaves its mother, and it is clear that whilst holding on in this

FIG. 87.—The arm of *Hapalemur guseus*, showing its brachial mammæ.

way it would be difficult for it to use the inguinal teats, if any existed, whilst it is within easy access of the pectoral pair. It seems to me reasonable to suppose that this habit of carrying the young has led to the pectoral teats being most used, and their gradual enlargement from increased use would slowly bring about the suppression of the little used inguinal set. Usually the nipples are present in the male, and have a disposition corresponding to that of the female.

A review of the arrangement of the milk glands in mammals generally indicates that primitively they were arranged in two rows, extending along the ventral aspect of the trunk in both sexes ; and in situation they correspond to the course of the deep and superior epigastric arteries, an extraordinary anastomosis effect-

FIG. 88.—*Lemur macaco* and young. (After Sclater.)

ing a free communication between the subclavian arteries at the anterior part of the thorax and the iliac system of arteries at the groin. This vascular arrangement is peculiar to mammals, and has direct relation with the mammary glands.

When the mammæ, or nipples, are increased beyond the number normal to a given mammal, they are said to be accessory or supernumerary, and occur in three forms :—

1. As nipples or teats.
2. As mammæ, furnished with teats.
3. As nipple-less mammæ.

Such mammæ, or nipples, may be due to atavism, or arise by dichotomy of a normal gland, or occur as a spontaneous variation or " sport."

Notwithstanding the scanty evidence at the disposal of Darwin when he wrote " The Descent of Man," he regarded supernumerary mammary glands as reversions, although he considered the opinion weakened by their occurrence on the thigh and back. Since then our knowledge on this subject has increased, and has had the effect of strengthening Darwin's opinion. We will first discuss those examples which may be safely regarded as atavistic. From what has been already stated regarding the ventral disposition of these glands, their association with the deep epigastric arteries, and subsequent suppression in the pectoral or abdominal regions, it would be necessarily inferred that accessory mammæ, or nipples, would reappear in such situations most frequently; this is precisely the case. So frequently are supernumerary nipples present in the human subject that Dr. Mitchell Bruce in three years saw sixty-five examples. Among 207 men examined consecutively, nine per cent. had an extra nipple, and of 104 women, four per cent. These observations have induced other observers to look into the question, with a confirmatory result; my own experience as to their frequency coincides with the above statement. As a rule only one extra nipple is found, but occasionally two may be detected. A typical specimen is given in fig. 89: the nipples are situated exactly in the line of the deep epigastric arteries.

As accessory glands, or nipples, are so frequent in the human subject, it occurred to me that they ought

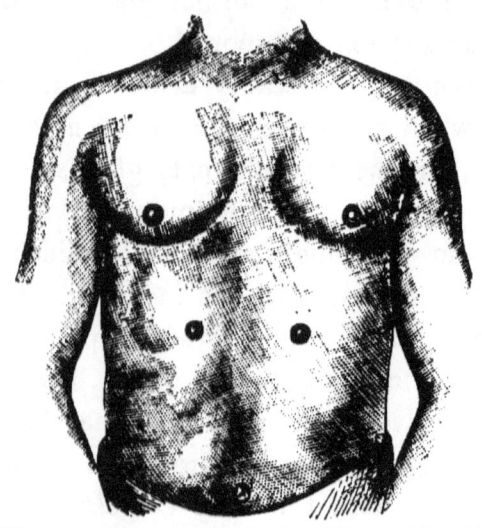

FIG. 89.—Two supernumerary nipples in a Man. (After Lichtenstern.)

theoretically to occur in quadrumana. To this end I examined systematically all monkeys coming under my observation, and in a short time secured two well-marked examples. The first was a female macaque (*Macacus sinicus*) (fig. 90). On the left side, about an inch below the normal gland, an accessory and well-formed nipple was detected; in size it equalled the normal teats, and was associated with glandular tissue and traversed by ducts. The second monkey came to hand a few months later. It was a male *Cercopithecus patas*: in this case the nipple occupied a similar position to that in the macaque; it was unassociated with glandular tissue. The parts are preserved in the museum of the Royal College of Surgeons, England.

Among domesticated mammals, cows present one,

FIG. 90.—A female Patas Monkey (*Cercopithecus patas*) with a supernumerary nipple.

sometimes two, and even three accessory teats; such extra teats are always situated posteriorly to the normal ones. In some cases they transmit milk, and cows have been known to furnish milk from seven teats. An instance has been reported to me, from a reliable source, of a cow which had two rudimentary teats in addition to the normal four. In consequence of an injury one of the normal teats became obstructed; this accident was compensated, as one of the accessory teats enlarged and was regularly milked. A female zebu (*Bos indicus*) at the Zoological Gardens has an extra well-formed teat on the right side. Goats and sheep are normally furnished with two teats, but it is not unusual to find a rudimentary pair posterior, and occasionally anterior, to the normal teats.

We may now consider supernumerary mammæ occurring in the line of the deep epigastric artery which are not atavistic. In the human female, and occasionally in the cow, we find mammary glands with bifid nipples; in other cases we find two mammæ coalesced, but each has a separate teat. These conditions are not reversions, but arise by dichotomy from a single gland germ. Intermediate stages between bifid nipples and two separate mammæ have been recorded.

Aberrant mammary glands, or those arising as spontaneous variations, claim careful consideration, for they may arise on any cutaneous surface which is rich in sebaceous glands: as a rule they are furnished with nipples, but not constantly. Among the rarer positions for such aberrant mammæ are the thigh, shoulder, and cheek. The oft-quoted case of a functional mamma on the thigh rests upon the authority of a committee of the French Academy of Sciences. Barth recently gave an account of a rudimentary mamma connected with the skin on the cheek of a maiden aged twenty years, it was situated immediately in front of the right ear, and had an areola of pigmented skin; when removed, and the parts examined microscopically, glandular tissue was found beneath it.

The most frequent situation on the exterior of the body in which to find aberrant mammæ is the arm-pit; here they appear as glandular masses, communicating with the free surface of the skin by means of nippleless pores, and in parturient women furnish milk and colostrum. Champneys has particularly investigated this form of mammæ, and finds that they are of frequent occurrence. Axillary mammæ with nipples have been recorded by several reliable observers.

The tendency of protected cutaneous surfaces to become glandular is a matter of great interest, not only in connection with aberrant mammæ, for it serves to explain the abundance of teats lodged within the pouch of opossums.

It is a noteworthy fact that cutaneous and mucous recesses are as a rule richly supplied with glands. Take, for instance, the large glands besetting the ocular aspect of the eyelids, the crowd of glands lodged in the armpit, in the external ear-passage which secrete ear-wax, and in many situations which readily suggest themselves to anatomists. As far as I have looked into the matter, no one has suggested any useful purpose served by glands in these situations worthy a moment's reflection.

The consideration of cutaneous recesses leads us to deal with the remarkable pouch of marsupials, which is neatly described as *the bag of the opossum* by Paley in his "Natural Theology":—"A false skin under the belly of the animal forms a pouch into which the young litter are received at their birth ; where they have an easy and constant access to the teats ; in which they are transported from place to place ; where they are at liberty to run in and out ; and where they find a refuge from surprise and danger."

On the wall of this pouch formed by the mother's belly we find nipples and milk glands ; in kangaroos and phalangers four teats are usually present ; in opossums they are more numerous. *Didelphys virginiana* has six on each side and one in the middle ; the crab-eating opossum (*Didelphys cancrivora*) has eleven teats, the odd one occupying the centre, whilst the remaining ten are disposed in a circle around it.

Such luxuriance of nipples in so limited a space is unknown in mammals outside the order marsupialia, and in them is probably due to the protecting influence of the pouch, as the following evidence indicates :—

Malkmus has recently drawn attention to some points of interest in the structure of the walls of the cutaneous recesses which exist in the inguinal region of sheep, or, as he terms them, rudimentary marsupia. On examining the inner aspect of a sheep's flank we find it smeared with an unctuous material not unlike cerumen, or car-wax. In the fold of each groin, close beside the dugs, a shallow recess is seen (fig. 91) usually containing a quantity of the same brown, unctuous material which besmears the flanks, and, as far as my observations extend, it is more abundant when the ewe has a lamb by its side. The walls of this recess are almost devoid of wool, but are beset with a number of large sebaceous glands, thus furnishing an excellent example of large and numerous skin glands occurring in a protected situation.

It has long been established that milk-glands are modified sebaceous glands, and the difference in the

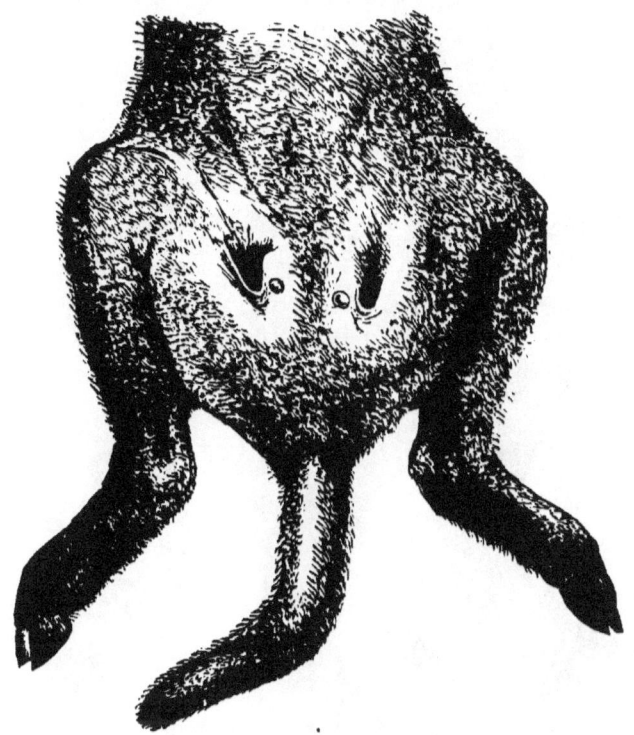

Fig. 91.—The inguinal region of a Lamb, showing the cutaneous recesses in its groin.

nature of the secretion is not very great. The sebaceous variety furnishes a fatty unctuous substance; mammæ, a natural emulsion known as milk, consisting of oil globules suspended in fluid, containing various salts and a proteid body, casein, with a small but variable proportion of albumen.

174 EVOLUTION AND DISEASE.

As all cutaneous recesses exhibit a marked tendency to develop glands, the large size and abundance of the mammary glands within the marsupium of opossums receives an explanation.

A comparison of the inguinal pouches in the sheep

FIG. 92.—The marsupium of a Phalanger (*Phalangista orientalis*) opened to show the number of teats.

with the marsupium of the phalangers, &c., is instructive, as it indicates in a significant manner the mode by which a specialized bag, or marsupium, may have originated, besides throwing light on the number of teats it contains.

By far the most remarkable situation in which milk-glands arise, is in the interior of some peculiar skin-containing tumours of the ovary. I have traced every stage from a sebaceous gland to a nippled mamma, furnishing milk, in such tumours—a situation in which it would puzzle even Paley to imagine a useful purpose.

These facts indicate clearly enough that a mamma may make its appearance in any situation of the body if the favouring conditions exist—*i.e.*, a cutaneous recess abundantly supplied with sebaceous glands, and preferably on the ventral or protected aspect of the body; warmth and moisture appearing to favour their development.

The careful consideration of such cases clearly shows that to attribute such aberrant mammæ to atavism is inconsistent with sound reasoning: it is, however, obviously erroneous to regard all forms of supernumerary mammæ and nipples as "sports," but I am firmly of opinion that only those occurring in the line of the deep epigastric arteries, excepting those clearly due to dichotomy, can be with any justness considered as reversionary. In the section on tumours we shall have to return to this subject, especially in reference to cancer.

CHAPTER VIII.

THE TRANSMISSION OF MALFORMATIONS AND ACQUIRED DEFECTS.

IN this chapter it is not proposed to discuss in all its bearings the question of heredity, or even to propose a theory to account for the transmission of defects from parents to offspring. At the present day such an undertaking is rendered unnecessary, for all thinking persons are unanimous in believing that malformations of development, as well as the tendency to certain diseases, are inherited, or, as it is familiarly expressed, "run in families." If anything, perhaps there is a disposition to ascribe too much to inheritance instead of submitting suspected cases to critical analysis. My object on the present occasion is to criticize a few examples of supposed instances of the transmission of characters the result of mutilation.

I shall commence with an easily observed part of a mammal's body—the ear, or pinna. In the chapter on Environment it was pointed out that good evidence leads us to believe that it is justifiable to regard the pinna as an enlarged operculum, modified for acoustic purposes in terrestrial mammals. Before discussing the malformations of the external ear we may profitably consider some facts connected with its anatomy.

TRANSMISSION OF MALFORMATIONS.

The various projections and depressions of the human pinna have received at the hands of anatomists distinguishing names; with these we must make ourselves acquainted. In the drawing of the pinna (fig. 93), H indicates the helix, and the prominence X is the antihelix; below the helix we recognize a projection T known as the tragus. In elderly persons this is furnished with some stiff hairs, sometimes of great length; opposite the tragus we note another elevation marked A in the sketch, this is the antitragus. The tragus and antitragus are separated by a deep notch, the intertragic fissure. The lower boundary of this fissure is formed by the lobule L, which varies considerably in size in different individuals. The significance of the terms helix and antihelix is easily understood, but the meaning of the term tragus, from the Greek, signifying a goat, is not so evident. Hyrtl suggests the following explanation:—In elderly persons the tragus not infrequently has some stiff hairs growing from it, known as goat's hair. The presence of these hairs was formerly regarded as indicating a sensual individual. It is probable that such ears, when the hair on the tragus was abundant, caused them to resemble the sharp-pointed ears of the goat-footed satyrs (ægipans). In older works

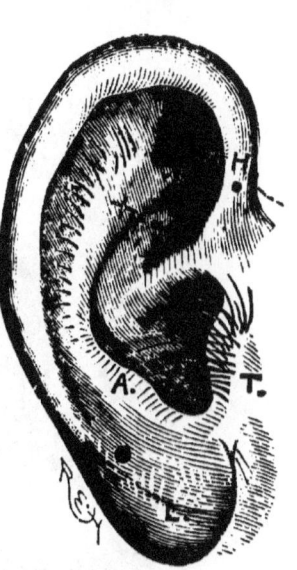

FIG. 93.—The human pinna; H, helix; X, antihelix; T, tragus; A, antitragus; L, lobule.

on human anatomy the tragus is referred to as *hircus* or *tragus* (Latin *hircus* = goat), on account of the hairs growing from it.

It is unnecessary for me to enter into the question of the vestigial nature of man's pinna; this has been again and again pointed out and admits of no doubt. (Like

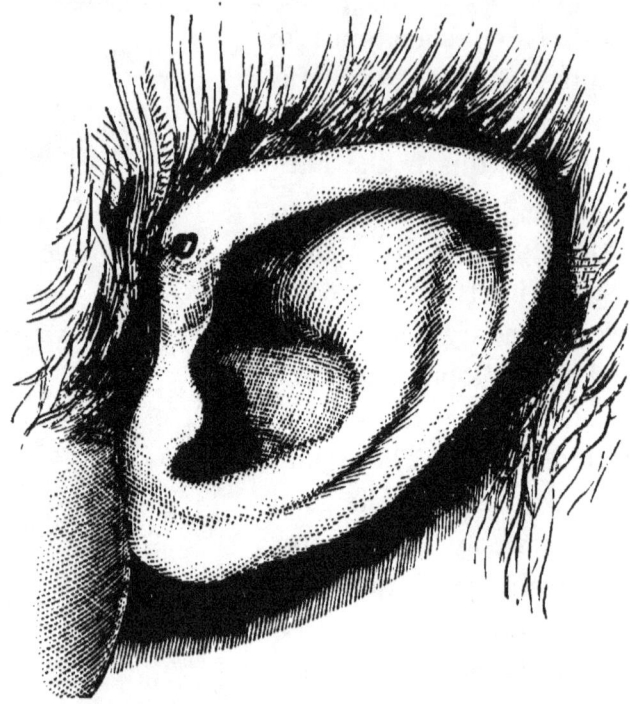

Fig. 94.—A malformed auricle with a sinus in the helix. (After Paget.)

vestigial organs in general it is very variable. The pinna may be absent, is often malformed, presents accessory parts and not unfrequently is occupied by some congenital fistulæ and recesses.

It is by no means uncommon to find in the helix, at the spot marked 4 in fig. 93, a small sinus, or recess, capable of admitting a probe sometimes to the extent of an inch, or even more. Such recesses are known as congenital auricular sinuses; they are furnished with glands which secrete an unctuous material resembling ear-wax, often troublesome to the individual on account of its amount. These sinuses are usually associated with malformed ears; for a typical example see fig. 94. They are hereditary, and often appear in several members of a family. Congenital fistulæ occur in the lobule, but they are very rare, only two or three examples have been placed on record. One excellent specimen has come under my observation. In this case the fistula was so complete that a few days after birth an earring was suspended in it. Only one pinna was affected. The situation of the fistula in this case is indicated in the lobule of the pinna, represented in fig. 93.

If any one possesses the patience to examine a large number of ears he will find that it is by no means uncommon to see a shallow furrow separating the lobule from the antitragus, and terminating occasionally in a slightly marked notch on the posterior rim of the auricle. The direction of this furrow varies greatly; sometimes it is nearly vertical, at other times oblique, and in rarer instances is so pronounced that it seems to divide the lobule in two parts. Among other variations exhibited by the pinna is the presence of a supernumerary tragus, which may or may not be associated with a well-shapen auricle. Two common forms are shown in the accompanying drawings (fig 95). These so-called supernumerary auricles are probably due to dichotomy of the tubercle which gives rise to the tragus.

In order to appreciate fully the bearing of these facts and some others which will be mentioned, it will be necessary to take into consideration the chief points connected with the development of the pinna as an illustration of the care requisite in interpreting variations and abnormalities from the evolutionary point of view. The details of the development of the human pinna have been carefully described by Professor His. At the end of the first month of embryonic life, the first branchial cleft is surrounded by six rounded, slightly prominent tubercles,

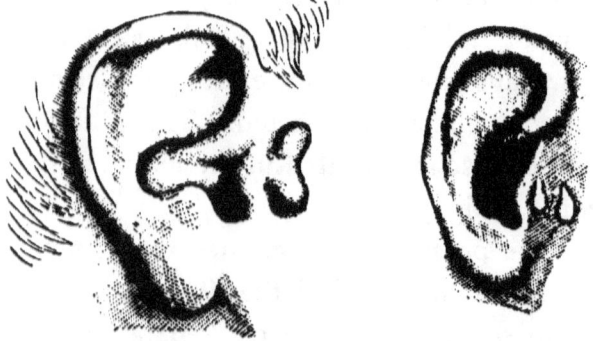

FIG. 95.—Human Ears with so-called supernumerary auricles.

numbered I to VI in fig. 96, and it is by the coalescence of these six tubercles that the pinna is formed. The details of their fusion are briefly these:—tubercles I and V unite across the fissure and form tragus and antitragus, the gap, or notch, between them is represented in the adult as the fissura intertragicum; tubercle VI unites with these to form the lobule; tubercle II forms the helix, whilst III lengthens out and forms the conchal rim and the tail, or cauda helicis, whilst the tubercle marked IV becomes the antihelix.

These facts explain readily enough the mode of origin

of the various congenital defects of the pinna. Thus a fistula in the helix results from incomplete coalescence of the tubercles II and III, whilst a fistula in the lobule represents a gap between the tubercles I, V, and VI. The furrow in the lobule is a fault where the lobule comes into relation with the tail of the helix. Should any of

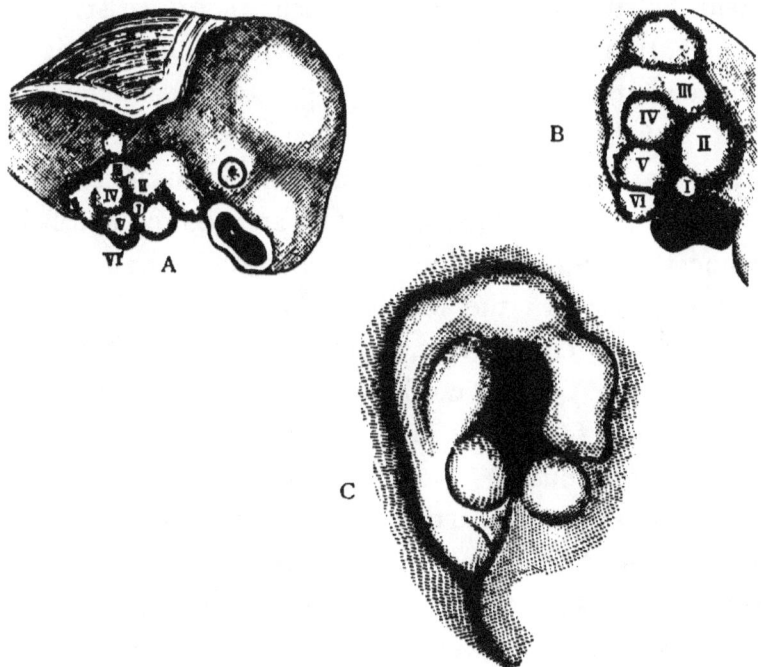

FIG. 96.—A, the six tubercles surrounding the first cleft; B, an intermediate stage in the coalescence of the tubercles. C, a still later stage. (After His.)

the tubercles fuse superficially and leave an intermediate space, this may subsequently dilate and form what is known as a dermoid cyst—that is, a tumour with a central cavity lined with skin resembling that covering the pinna.

These facts are of some importance because, with an

imperfect knowledge of the mode of origin of the pinna, these conditions might admit of a somewhat different interpretation. For example, the congenital hole in the lobule could on superficial examination be interpreted as the result of piercing the lobe of the ear for the suspension of ornaments, but the details of the development of the pinna offer a much more satisfactory explanation of the phenomenon. A careful analysis of the facts at our disposal strongly points to the conclusion that defects due to mutilations are not inherited. Recently Dr. Emil Schmidt, of Leipzig, has carefully argued a case in which he thinks an acquired defect in connection with the pinna was inherited; the facts are briefly these:—In the left ear of a child he noticed a peculiar congenital fault, consisting of a cleft in the lobule (fig. 97). The mother also had a similar defect in the lobule of the left ear; in the case of the mother the cleft was due to an injury. When eight years old, whilst playing, the earring was torn through the lobe. The edges of the wound did not completely unite, and the lobe was subsequently pierced to restore symmetry. Thirteen years afterwards she married, and gave birth between 1860-73 to eight children. The second child—a boy—exhibited the peculiar defect described above: the remaining children possess pinnæ with normal lobules. A comparison of the ear of mother and son shows that in the son's ear the cleft is situated more posteriorly and higher than in that of the mother.

Dr. Schmidt, who is well acquainted with the details of the mode of formation of the pinna, discusses it at some length in connection with this case with very great care,

TRANSMISSION OF MALFORMATIONS. 183

and in the end states that he is inclined to regard this as an example of the inheritance of acquired defect, for the following reasons:—The limits between the lobule and the tail of the helix are never so sharply defined as in this case. The cleft is situated too posteriorly to correspond to the line of fusion of the fifth and sixth tubercles, and the tubercle for the antitragus is, in the embryo, some distance from the external margin of the ear.

It is unnecessary to discuss the reasons advanced by

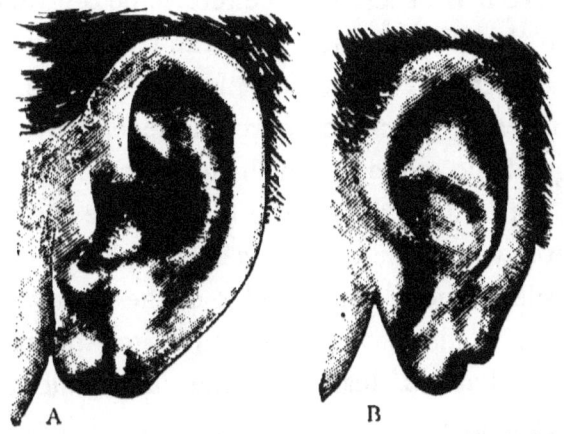

FIG. 97.—A, pinna of Mother showing the rent in the lobule where the earring was torn away ; B, pinna of Son with congenital defect. (After Schmidt.)

Schmidt in support of this view, for they are nullified by some observations made on this case by Professor His, in a subsequent number of the *Correspondenz-Blatt für Anthropologie* (March, 1889), to the effect that the defect in the ear of the son not only differs from that in the mother in its general character, but occupies a different position in the lobule, as is easily seen in comparing the two pinnæ.

The slit, marked F in the sketch, in the mother's pinna was made subsequent to the accident, so as to enable her to wear an ear ornament.

Reference to the drawings in fig. 96, illustrating the development of the pinna, would show that the defect in the son may be reasonably attributed to feeble development and incomplete union of the tubercles for the lobule and antitragus.

The facts relative to malformations of the pinna have been considered in detail, for they serve to indicate the extreme care it is necessary to exercise before we regard a defect in the offspring as due to the transmission of an acquired defect in the parent. In Schmidt's example, the conclusion at which he arrived, independently of our knowledge of the mode by which the pinna is developed, is clearly not in accord with the facts of the case, but in the example where the child was born with a perforation in the lobule, without a knowledge of the embryology of the pinna, it would have seemed unreasonable to doubt but that we had to deal with the inheritance of an acquired defect.

The facts considered in relation to acquired defects of the pinna are of importance in connection with the subject of the transmission of mutilations. It appears that, immediately on the announcement of the evolution theory, those who were antagonistic urged that the practice of docking the tails of such animals as horses and sheep for so many generations should have produced animals with short tails. The Hebrew custom of circumcision is as necessary now as in the time of Moses. To these may be added the habit of piercing the lobe of the ear, practised early in life, and in most nations of the world, and even in tribes the most barbaric, yet there is not the slightest evidence that the result of such mutilations are transmitted, and the ear is of all parts of the body the most easily and frequently observed. If such injuries—for so we must regard them—were transmitted, we ought to find the offspring of wild birds in confinement, which have been pinioned, lack the bastard wing; such is not the case, nor are they in any way malformed in consequence of the mutilation to which the parents were subjected. It is inconsistent with the fundamental principles of evolution that the effects of removal or injury to parts of the body should be transmitted to the offspring. It may perhaps be interesting to briefly mention two instances which, in the absence of more accurate information, have been interpreted as due to inheritance of acquired conditions.

The first instance concerns the tail feathers of *Momotus;* this bird has the singular habit of picking away the web of the central feathers of its tail until they assume a spatulate condition. In consequence of the constant stripping of the web the feathers, when they first appear, are naturally narrower in the places where they are habitually denuded. The habit is practised by both sexes alike. The matter has been carefully investigated by Mr. Salvin, who is satisfied that this plucking process is practised by the motmots.

These observations are interesting, for it is thought probable that the spatulate or racket-shaped tails in some other birds may have been brought about in some such way, so as to become a permanent condition. All degrees may be met with, from a broad web to others in which the rachis of the tail feathers is clean from the first.

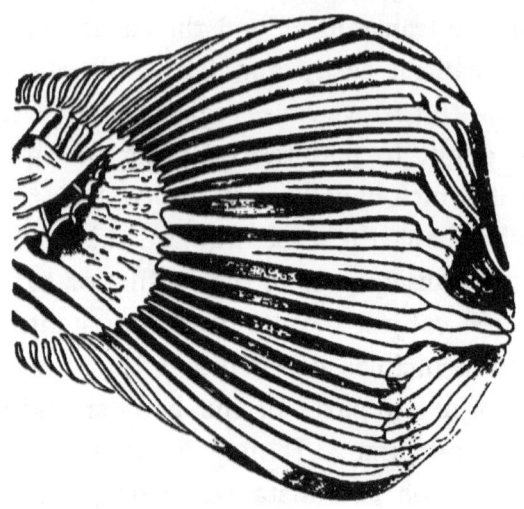

FIG. 98.—The tail of the so-called Tailless Trout of Islay.

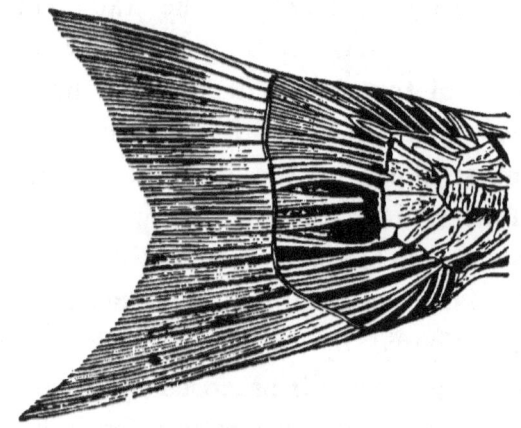

FIG. 99.—Tail of a normal Trout for comparison. (After Traquair.)

The second case is furnished by the so-called tailless trout of Islay. Professor Traquair has published an interesting account of the malformation of the tail in these fish. The most salient peculiarity of these Lochnamaorachan trout is that the rays of the caudal fin are abnormally shortened, coarse at the extremities, and deficient as to the amount of dichotomization and number of the transverse joints; besides which, they also show a tendency to coalesce at their terminations. By the convergence downwards of the upper long rays, and upwards of the lower ones, the fin assumes a rounded form instead of presenting the usual broad fan-shaped aspect. The abnormal condition of the extremities of the rays may affect other fins besides the caudal (fig. 98).

The lake in question is about 1,000 feet above the level of the sea. It is about an acre in extent, and so shallow that a man can wade through it; the bottom is of quartz rock. Several other lochs near contain trout, but none are "tailless." So constant is this abnormality in trout taken from the lake in question, that one keen fisher, with thirty years' experience of this loch, has never taken any but docked ones.

It is quite possible, indeed probable, that a more perfect knowledge of the mode of development of the spatulate or racket-shaped feathers in the motmots, and of the tails in the Lochnamaorachan trout would put a new complexion on the matter. It may be confidently stated that at present there is no satisfactory case known of the transmission of a defect, the result of mutilation.

Inherited Malformations.—There can be no reasonable doubt that defects arising in the course of development of an individual are in many instances transmitted to the offspring. Malformations, as such defects are called, arising in this manner are usually classed as arrests of development, and as it is clearly established that the embryological history of a complex individual may be regarded as an abbreviated history of its evolution, it necessarily follows that, should the development of a part be arrested at any particular stage, we should expect to find, in some less specialized mammal, this stage represented as a permanent condition. For instance, children are often born with club feet; the commonest congenital form is that which in the sole of the foot looks inwards and upwards, and the heel is slightly raised (talipes equino-varus). In the child before it is born the feet are for several months in this position, and gradually pass into that assumed normally in the adult. Not infrequently the foot fails to assume a position at right angles to the leg, and malformation is the result. Now it is very instructive to remember that the orang's foot is in the position typical of talipes equino-varus, and this is not limited merely to the position of the foot, but extends also to the disposition of the articular surfaces on the ankle-bone.

It is far from my intention to enter in a detailed manner into all forms of malformation which may be transmitted; that such characters are inherited is indisputable, but it may be useful to describe an instance in which it has been possible to produce a permanent variety in which the existence of a malformation is the distinguishing feature. This is the more important because as a rule defects of this class appear sporadically in a family, but do not reappear so constantly to become as it were, a matter of entail. This may be in part explained, perhaps, by the fact that attempts are not made to propagate the malformation, as is done in domesticated animals.

The frequent and familiar malformation known as hare-lip and cleft palate is not by any means confined to the human species, but occurs in horses, calves, sheep, dogs, and even lions. In the human subject it has been

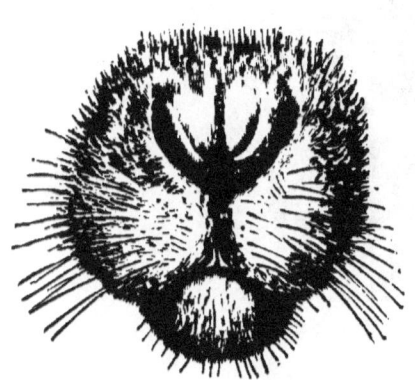

FIG. 100.—The nose and lip of a Hare, showing the cleft.

FIG. 101.—The nose of a Dog, showing the median furrow.

known to affect several members of the same family, and to occur in the offspring of the affected members. The defect takes the name hare-lip because the hare, in common with a few other mammals, exhibits a median cleft in its upper lip. This cleft in the hare is an exaggeration of the furrow which is present on the upper lip of many mammals, and extends on to the nose (figs. 100 and 101). In children affected with hare-lip the cleft is rarely situated in the middle line, but to one or other side;

sometimes the lip is cleft on each side, the median piece standing forward from the nose like a peninsula. An inquiry into specimens of hare-lip occurring in mammals generally, discloses the fact that in some the cleft may be lateral, as in man, but in others it is median in position. The following facts are interesting in this direction :—

In the summer of 1886, whilst staying in Paris with Mr. H. W. Freeman, of Bath, we procured a pug-bitch

FIG. 102.—Cleft-lip and nose in a Dog.

with a peculiar cleft in its nose. At first we thought that the case was an example of hare-lip, but on making inquiry we found it to be a distinguishing feature of this breed of dogs and that it conferred upon them a high money value. The bitch was brought to Bath, and Mr. Freeman was successful in crossing her with a Skye-terrier and obtained some pups. Half the litter had normal noses like the dog, the remainder had split noses and lips like the mother. The pups from

TRANSMISSION OF MALFORMATIONS. 191

this litter have since had young with cleft noses, and the malformation seems to be well established. The deformity consists of a median vertical split in the upper lip, extending some distance between the nostrils and involving the hard palate (fig. 102). The extent to which the bony portion of the palate in-

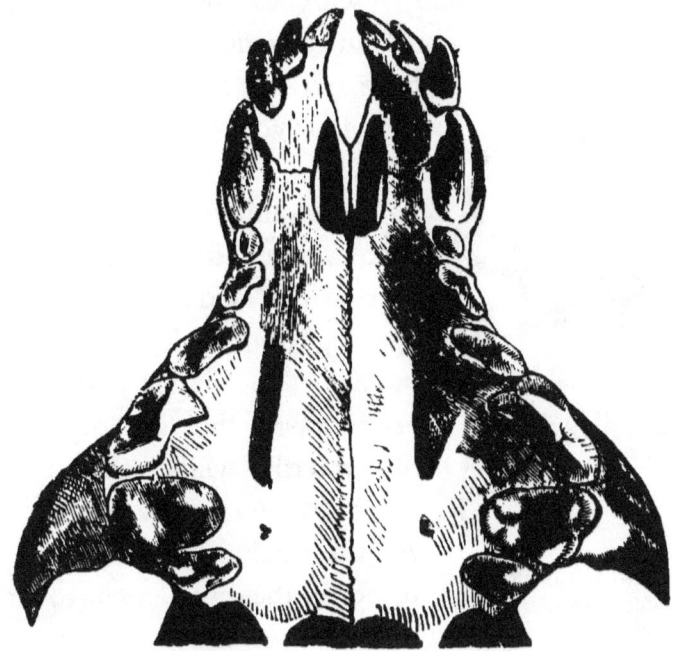

FIG. 103.—The hard palate from a Dog with hare-lip, showing the extent to which the hard palate is involved.

volved may be seen in fig. 103. Apart from its special interest the malformation is of value as showing that defects, arising from arrested growth during embryonic life, are transmitted, and this is clearly the case not only in connection with the lips, but with the pinna, eye, eyelids, and many other parts of the body.

It now becomes necessary to detail briefly the leading

facts connected with the development of the mouth in order to show that we have here to deal with an arrest of development, as well as to show the reason why the cleft in the dog's nose is median, whereas in man and several other mammals, in which the deformity has been recognized, the split is, as a rule, lateral. Fortunately the admirable researches of Professor His remove much of the difficulty which would otherwise have existed, and the account of this embryologist will be closely followed.

The mouth of the human embryo at the fifth week of intrauterine life is represented by an opening from which five fissures radiate. The upper pair are the orbito-nasal, the two lower form the mouth, whilst the median fissure separates the lower jaws. As the median process develops to form the nose, two rounded prominences make their appearance at each angle. These will be referred to as the globular processes (fig. 104); these processes furnish the alæ of the nose and the intermaxillæ; later they are joined by the lateral pieces to complete the lip.

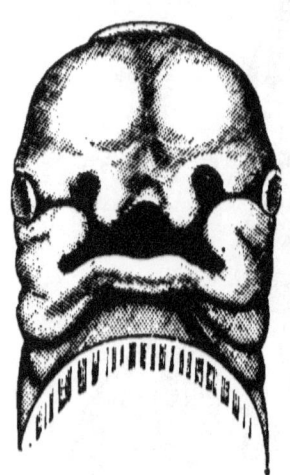

FIG. 104.—Human embryo of the fifth week showing the parts out of which the lips, mouth, and nose are formed.

In some mammals, especially rodents, the globular processes fuse with the lateral pieces but do not fuse with each other and remain permanently separated, thus explaining the occurrence of a persistent median cleft in the hare. Thus the cleft nose in the pug (fig. 101) is due to the non-union of the globular processes. It is a fact of some interest that in the embryos of some kangaroos (*Macropus*) the fissures in the lips persist longer than in any other mammals I have yet had an opportunity of examining.

These facts concerning the transmissibility of defects in the dog's nose tend to show that it is probable, if it were possible to practise selective breeding in the human species as among dogs, a race of men with hare-lips and cleft-palates could be produced, and this view is further strengthened by the knowledge that in some districts where the inhabitants are not very numerous a sort of indirect selection goes on with the effect of perpetuating deformities.

An excellent example of this is given by Professor Bertram Windle in a paper on "Congenital Malformations and Heredity." He remarks that a singular account, given by Devay on the authority of Dr. A. Potton, seems to show that under favourable circumstances, continued for a sufficient length of time, a separate breed of men, possessed of some malformation, might be produced, whilst it also shows that the instability of such a condition, which must persist for some time, leads to its being easily stamped out by the introduction of fresh and untainted blood.

In the department of Isère not far from Côte-Saint-André and Rives, there is a small isolated village called Izeaux, lost, so to speak, in the midst of a plain, called the plain of Bièvre, which, if not completely uncultivated, was at least very barren. The roads were difficult, if not impracticable. The inhabitants, thus cut off from the outside world, intermarried very freely. At the end of the last century sexdigitism, both of the hands and feet, suddenly appeared, and in thirty-five or forty years almost the entire population was thus affected. When, in 1829 and 1836, says M. Potton, I observed this strange phenomenon it was present in some subjects only in a very rudimentary manner; amongst many it was only a large tubercle containing a hard osseous body, and fixed to the side of the thumb, a more or less well-formed nail terminating it. At this time the influence of crossing, due to the opening-up of communications, was making itself felt. In 1847 I had occasion to see a foreman, originally from this locality, who married and settled in Lyons. He was affected with the malformation, but was the father of four normal children. At the time of writing, he goes on to say, the anomaly has almost completely disappeared from the district.

This isolation of villages helps to explain the endemic cretinism of Alpine countries. It has been shown, on reliable authority, that cretins are most abundant in villages where intercommunication with towns or neighbouring villages is difficult, so that the inhabitants intermarry freely; since the introduction of railways freer communications have been opened up, and new blood introduced in the affected districts; this has had the effect of diminishing the number of cretins.

This condition of things is illustrated by the following example :—There is preserved in the museum of the Royal College of Surgeons a fish with a large tumour growing from its side. It was suggested that the pond

TRANSMISSION OF MALFORMATIONS. 195

from which this fish was taken should be dragged for the purpose of ascertaining the existence of other fish similarly affected. When this was done several fish of the same species were taken with tumours of the same kind growing upon them.

In this case it is reasonable to suppose, in the absence of evidence to the contrary, that the inhabitants of the pond being limited, the chances of this defect being transmitted were greatly increased. This view is equally applicable to the case of the tailless trout of Islay.

The curious condition of the skeleton of the fish *Chætodon*, described by William Bell in the "Philosophical Transaction," 1793, deserves mention in connection with this subject. He writes:— " The fish is frequently caught at Bencoolen and several other parts on the west coast of Sumatra. The skeleton is very singular, many of the bones having tumours, which in the first fish Mr. Bell saw he supposed to be exostoses arising from disease, but on dissecting a second found the corresponding bones had exactly similar tumours, and the fishermen informed

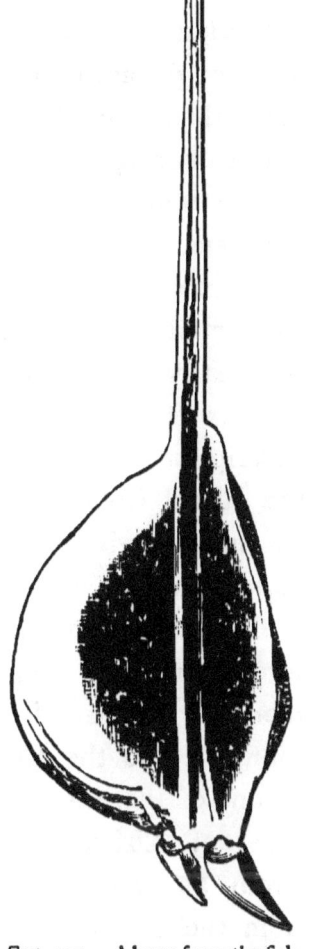

FIG. 105.—A bone from the fish Chætodon with a tumour, shown in section, attached to it. (Nat. size.)

him that they were always in this fish; he therefore concluded them to be natural to it.

Bones of this kind are not uncommon in osteological collections. Cuvier explains this by stating that they are brought home as curiosities by travellers who have eaten these fish. No one has attempted to explain these curious bones, which are very characteristic; at one end they present a tumour about the size of a chestnut, very hard, smooth, and as dense as ivory. Articulating with the tumour by means of a shackle-joint, is one, often two, small rays. On section the outline of the ray can be clearly defined running through the midst of the tumour (fig. 105). The fish on which Bell's original description was founded is preserved in the museum of the Royal College of Surgeons, and a drawing of the specimen is given in "The Transactions of the Pathological Society of London," vol. xxxix.

A consideration of the bones of the *Chætodon* is of interest in connection with what are known as the horned men of Africa. In 1883 Professor Macalister communicated to the Royal Irish Academy a photograph of one of these men, who came from Akim on the west coast of Africa (6° N. latitude and 1° E. longitude). From an examination of the photograph, as well as from the descriptions of those who had examined the man, the so-called horns appear to be outgrowths from the malar bone and nasal process of the upper jaw. This is by no means an infrequent situation for exostoses in Europeans, and, as a rule, such outgrowths are symmetrical, and give a very hideous appearance to the individual so unfortunate as to possess them. Such bony outgrowths are believed by many surgeons to be hereditary.

In his concluding remarks Macalister says, "That outgrowths here may be really race characters is not to be entirely ridiculed, for the neighbouring malar bone which here, according to O'Reilly's description, participates in the swelling, certainly shows certain race peculiarities, such as the bigger *Tuberositas malaris* of the

FIG. 106.—A so-called Horned Man of Africa. (After Lamprey.)

Mongolians, and the *Processus marginalis*, whose race peculiarities have been pointed out by Werfer."

Some further particulars relative to horned men have been furnished by J. J. Lamprey,[1] of the Army Medical Staff, who has had opportunities of examining carefully three persons from different localities in Western Africa, having peculiarities similar to Macalister's case. In each

[1] *British Medical Journal*, 1888.

instance there was a clearly defined bony prominence over the infraorbital ridge of each maxillary bone; the skin in each case was freely movable over the abnormal outgrowth. The ages of the men were 38, 18, and 20 years. The prominences had existed as long as the men could remember, and caused them no inconvenience (fig. 106).

CHAPTER IX.

ANATOMICAL PECULIARITIES OF THE TEETH IN RELATION TO INJURY AND DISEASE.

MANY troubles and disasters arise in animals in consequence of injuries to their teeth, and injuries, trivial in one mammal, may in another be fraught with serious results, due to variations in the use and specialization of the dental organs.

We may commence this subject with the kangaroo. This mammal has three incisors in the upper jaw, but only one in the mandible. This lower incisor is procumbent and flattened from side to side; the outer surface is slightly convex, the inner flat, but has a median ridge; the margins of the tooth are sharp. The lower incisor, instead of antagonizing the upper teeth by means of its tip, or crown, meets them along its sharp outer margin (fig. 107). The lower incisor is provided with a large persistent pulp; the pulp chamber, contrary to what is usual in teeth, extends nearly to the tip. The points of these teeth, shaped something like a lancet, are exceedingly thin and brittle. As a consequence, the tips are easily broken, and if only a small piece is detached, the pulp is exposed (fig. 108). Kangaroos, like mammals of even high moral pretensions, have domestic differences, which occasionally lead to unpleasant consequences. In the encounter the tips of the incisors are broken, the exposed pulps inflame, suppurate, and lead to alveolar abscess, which in some cases terminates in death from absorption of septic matter. Occasionally the suppuration leads to extensive necrosis of the jaw. Death from

Fig. 107.—The incisors of a Kangaroo (*Macropus major*).

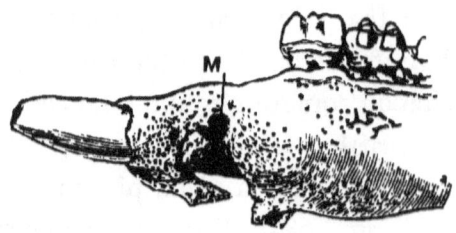

Fig. 108.—The anterior portion of a lower jaw of a Kangaroo showing the effects of alveolar abscess secondary to a suppurating pulp. M, mental foramen.

such a cause is not uncommon among kangaroos living in confinement.

In this case the danger depends upon the peculiar procumbent position of the lower incisor and the proximity of the pulp to the tip of the tooth. In the majority of mammals the teeth when fully formed cease to grow; in some—*e.g.*, the rodents—the teeth continue to increase in length throughout life: such teeth are said to possess persistent pulps. This is of great advantage to the animal, as it compensates for the continual wearing down of the tooth in consequence of the rough work to which such teeth are subjected. Under abnormal conditions a persistent pulp may be, and often is, of great disadvantage. If from accident to a tooth, or some injury to the jaw, the antagonism of such teeth be interfered with, the tooth or teeth which fail to antagonize cannot be used in mastication, and as their continual growth is not checked by wear, they may attain a length of several inches. In such cases the enlarged teeth may prevent the animal feeding, and thus bring about a fatal result from starvation; or the elongated tooth may re-enter the head or mouth, producing pain or even death, by piercing the brain. This last event is rare; the usual mode of death is from starvation—either the animal cannot bite its food, or the abnormality of the teeth prevents the mouth being opened. Such cases are exceedingly frequent in rabbits and rats; every gamekeeper of experience has met with many examples, and also those who keep white rats and rabbits as pets.

The study of such aberrant teeth is instructive, as it serves to throw light on the mode of origin of tusks. In many mammals it is common to find in the upper or lower jaw, sometimes in both, one tooth larger and more projecting than its fellows; this conspicuous tooth has a pointed extremity and is known as the canine. It is exceedingly well developed in most carnivorous mammals, lions, tigers, bears, cats, dogs, weasels, badgers, and the like. In those mammals with well-formed canines, the adjacent teeth, more especially those situated posteriorly to it in the dental series, are either small in size or absent in the adult animal. This reduction in size or number of teeth adjacent to those excessively enlarged occurs, not only in those animals with enlarged canines, but also when the incisors are unusually developed, as in the case of rodents, or rodent-like mammals as the aye-aye and the wombat. This happens so constantly that in all probability the enlargement of the incisors, or canines as the case may be, is directly responsible for this effect, partly by causing diversion of the blood-supply and partly from disuse. The combined effects of diversion of the blood-stream and disuse will be more fully illustrated in the case of tusks.

When a normal canine or incisor tooth is so long that it protrudes from the mouth when the lips are closed it is usual to term it a tusk. In most instances, as in the boar, walrus, elephant, and narwhal, the tusks protrude between the lips; in the babirussa the lower tusks protrude in this manner, but the upper pair make their way directly through the skin covering the upper jaw. Tusks, like the teeth of rodents, grow from persistent pulps, and many interesting pathological conditions arise in consequence of this peculiarity.

Among boars there is a tendency for the tusks to grow abnormally and describe circles. An example of this was described in 1733 by Cheselden;[1] the specimen is

[1] "Osteographia," London, 1733.

ANATOMICAL PECULIARITIES OF TEETH 203

preserved in the museum of the Royal College of Surgeons, England, and is figured below. In this specimen (fig. 109) the lower canines have grown excessively, turned backwards, and re-entered the mouth by piercing the integument and the jaw. The right tusk has tunnelled the bone for a distance of three inches and reappeared on the floor of the mouth and has described a complete circle. The left tusk, after re-entering the

FIG. 109.—Abnormal growth of the lower canines of a Boar.

mouth, seems to have crossed the buccal cavity so that its apex rests on the inner side of the right lower jaw. This is by no means an unique specimen, for on inquiry I have come across numerous examples of this aberrant growth of the lower canines of boars. The excellent museum of the Veterinary School at Alfort has a specimen resembling that of Cheselden. I have a canine of this character from a boar which measures thirty centimetres round the curve. In the museum of the Odontological Society of Great Britain the incisor of a hippopotamus (a huge pig) is preserved which has described a complete circle, the point of the tooth re-entering its own pulp chamber. The circle formed by this overgrown tooth has a diameter of forty centimetres.

The remarkable tusks of the babirussa, especially when the animal is confined in zoological gardens, are exceedingly prone to take an abnormal course, and instead of forming graceful curves beside the head, may

FIG. 110.—The head of a Babirussa. The upper canines are re-entering the skull.

deviate towards the middle line and enter the skull. Such an example taken from life is sketched in fig. 110; in order to prevent disaster the babirussa was thrown and ten centimetres of the tusks removed; they had, however, penetrated to the depth of twelve millimetres.

This deviation of the upper canines may be in part accounted for by the fact that, like tusks in general, they are slightly movable in their sockets, hence by rubbing them against the sides of the dens or cages, a false direction is impressed upon them. Careful consideration of such cases has induced me to believe that these aberrant tusks may be regarded as arising in the first instance as overgrowths, and that such malformations being frequently repeated the tendency has been transmitted to the offspring and eventually perpetuated as a normal character of the male.

The various steps of such a process may be studied in the pig tribe. In the wild boar the upper tusk projects about an inch from beneath the labial folds and curves slightly upwards; in the wart hogs it is much larger and may attain a length of several inches or more: in babirussa it is so strongly curved that instead of emerging from beneath the lips it directly pierces them. The transfixion of the skin covering the upper jaw by an abnormal tusk was illustrated in a striking way in the case of the celebrated African elephant Jumbo. During fits of bad temper Jumbo often damaged the tusks by contact with the walls of the den, and at last the pulp chambers became exposed, ending in alveolar abscesses of great magnitude. The constitutional disturbance caused by the suppuration would in all probability have ended fatally but for the undaunted bravery, skill, and ingenuity of Mr. A. D. Bartlett, who successfully attacked the elephant, and opened the abscesses through the cheek. It was through the incisions made for this purpose that the damaged tusks finally emerged, thus explaining why Jumbo's tusks projected through the cheeks instead of issuing from beneath the lips, as in elephants generally.

The tusks of elephants, from their large size, are liable to certain injuries which could scarcely occur in a smaller mammal. The centre of every growing tooth is occupied by a cavity known as the pulp chamber, which in the living tooth is filled with connective tissue, blood-vessels, nerves, and cells known as odontoblasts. The tissues constituting the pulp are actively engaged in forming dentine; in teeth and tusks with persistent pulps this pulp chamber is always relatively large, and as the tooth is worn at the apex the pulp adds new material at the base. In teeth with non-persistent pulps, like those of primates, carnivora, and the like, the pulp chamber diminishes with age, and in some the pulp becomes converted into bony substance known as osteo-dentine. When teeth become inflamed or carious the pulp often calcifies, and it is no unusual event to find in a carious tooth extensive ossification of the pulp in the neighbourhood of the carious cavity: this is salutary, for by this means suppuration of the pulp is prevented and often a tooth remains serviceable much longer than would otherwise be the case.

Ivory-turners have often found in the tusks of elephants such things as bullets, iron slugs, and spear-heads, yet on an attentive examination no trace of injury could be detected on the exterior of the tusk. These specimens attracted the attention of investigators such as Blumenbach, Haller, Cuvier, Goodsir, and Owen. The solution of the mystery was indicated by a study of the material surrounding the foreign body. Cuvier detected the irregularity of the dentine immediately surrounding the bullet, and Goodsir made an elaborate study of the whole question, and it became clear from the researches of Goodsir, Nasymth, and Owen, that this irregular

ANATOMICAL PECULIARITIES OF TEETH. 207

tissue is of the nature of secondary or osteo-dentine, and histologically agrees with that found in the pulp chamber of human teeth under abnormal conditions.

The presence of bullets may be explained in this way :—On reference to the section of the tusk in fig. 111

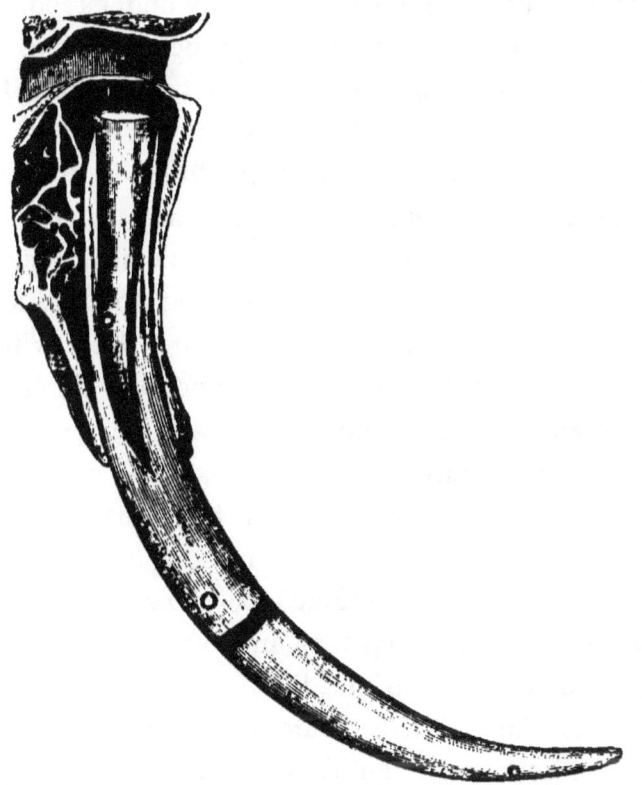

FIG. 111.—A diagram explanatory of the mode of encystment of bullets in elephants' tusks.

it will be seen that the walls of the pulp chamber are thin, hence a bullet fired at an elephant's head will readily enter the pulp; and should it lodge in this chamber inflammatory disturbance is the result, and the pulp immediately adjacent to the bullet forms irregular bone, or osteo-dentine, thus encysting the missile. In due course this foreign body moves onward with the segment of the tusk in which it is embedded, and eventually comes to lie in the exserted portion of the tusk until it is liberated by wear or the skill of the ivory-turner.

The entrance of a bullet is readily understood. The great force by which they are propelled carries them easily through animal structures, but with spear-heads it is rather different. Mr. Charles Tomes [1] gives an intelligible explanation of this condition in connection with the head of a spear embedded in an elephant's tusk preserved in the museum of the Odontological Society of Great Britain. It is to be presumed that a heavily-loaded spear was dropped by a native from a tree, with the intention of its entering the brain, upon the elephant as it was going to water. But in this case the spear penetrated the open base of the growing tusk, which looks almost vertically upwards, and then the iron point appears to have broken off. This did not destroy the pulp, but the tooth continued to grow, and the iron point, twenty centimetres long and four in width, became so completely enclosed that there was nothing upon the exterior of the tusk to indicate its presence.

The museums of the Royal College of Surgeons, England, and Odontological Society of Great Britain, contain many excellent specimens of foreign bodies embedded in the tusks and molar teeth of elephants, as well as masses of osteo-dentine removed from tusks apparently healthy. Osteo-dentine, in all respects similar to that formed in tusks under diseased con-

[1] "Dental Anatomy," 1st ed., p. 322.

ditions, makes up a large proportion of the tooth in some whales. This identity of tissue, arising in consequence of disease, in one animal with that which is normal in another, called forth the remark from Goodsir, that it "must be looked upon as another instance indicating the existence of a system of laws regulating the relations between healthy and morbid tissues."

In this connection a short reference to a curious affection of the teeth of whales may be made. In the grampus and sperm-whale the teeth consist of a hollow cone of dentine coated by cementum; the pulp chamber is very large and contains a quantity of osteo-dentine. In one species, *Orca*, Eschricht found a complete covering of enamel to the free portion of the tooth. In *Orca*, as in the porpoise, the teeth interdigitate when the mouth is closed, and as the teeth wear, owing to the friction, their broad bases come more into apposition. This wearing of the contact surfaces must, in the long run, inevitably open up the pulp chamber; as a rule it is prevented in consequence of the formation of osteo-dentine by the pulp. In some cases the development of osteo-dentine does not take place with sufficient rapidity, and the pulp becomes exposed and subsequently destroyed. A case of this nature has been minutely described by Eschricht and by Tomes.

No order of mammals exhibits so many deviations from the ordinary conditions of teeth as whales: not the least interesting abnormality in connection with our present purpose is the remarkable overgrowth of the teeth of *Mesoplodon*, secured by the naturalists of the *Challenger* Expedition, and carefully described by Professor Sir W. Turner. The illustration (fig. 112), which was prepared from a model of this specimen, represents the anterior part of the rostrum and lower jaw with two teeth of *Mesoplodon layardi*. This whale has only one tooth in each mandible (at least, as far as we know). In the present case the fangs of these teeth had become greatly elongated, and after emerging from the gum had curved backwards, upwards, and inwards, so as to cross each other on the dorsal surface of the whale's beak. The anterior border and inner surface of each tooth was

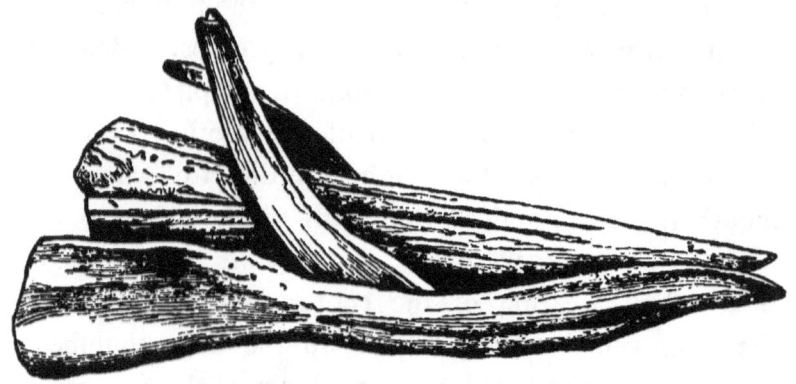

FIG. 112.—The rostrum of *Mesoplodon layardi*, showing the curved and elongated mandibular teeth.

smooth where they rubbed together during the movements of the mouth. The size of these teeth is as follows:—The extracted tooth measures 35 centimetres, 16 of which were embedded in the alveolus, or surrounded by the gum. The breadth of the tooth where it emerged from the alveolus was 9 centimetres. At the time this specimen was submitted to Turner, he was fortunate enough to receive a young specimen of Mesoplodon which enabled him to determine the nature of the various parts of this abnormal tooth, and he came to

ANATOMICAL PECULIARITIES OF TEETH. 211

the conclusion that its peculiar form was due to changes in the fang.

From mammals we may turn our attention to snakes. In toothed reptiles the teeth are attached to the jaw-bone in a manner differing from that which pertains to mammals. In the latter the teeth, as they develop, become surrounded by walls of loose porous bone, forming

FIG. 113.—Section through the jaw and tooth of a Python, showing bone of attachment.

an alveolus; when the teeth are erupted the roots or fangs remain embedded in this imperfect kind of bone; hence, when a tooth is extracted, a socket, or empty space remains, resembling that left by withdrawing a nail from a piece of wood into which it has been driven. On this account this mode of lodgment of teeth is termed gomphosis.

In snakes the teeth do not come directly in contact with the osseous tissue of the jaw; but, coincident with the development of the teeth, a formation of loose, bony matter occurs, whereby the base of each tooth is firmly cemented to the jaw-bone. Charles Tomes aptly describes this intermediate tissue as "bone of attachment"; it is of loose and open texture, resembling the alveolus. If from any cause this bone of attachment is softened or absorbed the teeth fall. It happens that snakes living in confinement are liable to a disease which induces softening of the bones of the skull, and as the bone which attaches the teeth to the jaw is, especially when compared with the bones of the skeleton generally—which are of almost ivory-like hardness—soft, and easily absorbed, the snake in consequence loses its teeth. The relation of tooth, jaw, and bone of attachment is represented in fig. 113. Premature shedding of teeth from this cause is not infrequent in snakes living in captivity; but I am unaware of its occurrence in snakes living wild.

Absorption of the alveolus occurs in mammals as age advances, leading to the edentulous condition of the gums which is one of the concomitants of advanced life; but the fall of teeth referred to in the case of the snake is due to constitutional defects, and not old age.

CHAPTER X.

CAUSES OF DISEASE.

Inflammation and Fever.—Thus far we have been concerned with diseases and malformations arising mainly from structural alterations due to the combined effects of variability and inheritance. We will now consider some examples of disease due to other causes.

Biological researches of the last few years have shown that many diseases of animals are due to the entrance, and subsequent multiplication in the system, of microscopic bodies known collectively as bacteria, or microorganisms. It has been clearly established by an overwhelming amount of evidence that such conditions as tuberculosis, glanders, actinomycosis and other contagious diseases are due to the action of these minute bodies.

For our purpose bacteria may be divided into two groups—those which when introduced into the body cause no harm, and those which produce disturbance either local or general. The latter are said to be pathogenic, or disease-producing. A study of the relation of bacteria to disease is of great interest, and sheds abundant light, not only on the nature of specific diseases, but also on the nature of a very remarkable condition known generally as inflammation.

Bacteria—using this term in a general sense—require high powers of the microscope and suitable methods of preparation for their satisfactory detection and identification: they appear as minute rods, or small rounded bodies; the latter are usually termed micrococci. In order to prove that any given bacterium is the cause of a particular disease, it is necessary to ascertain its constant presence in the blood or tissues of animals suffering from the disease, and then to cultivate it in media apart from the living body through several generations; on introducing the product of such cultivations into an animal, the disease should make its appearance and the micro-organisms occur in the lesions.

Pathogenic bacteria, when introduced into an organism either accidentally or designedly, give rise to trouble which may be either local or general. The local signs are peculiar, for the disease may in some cases be restricted to the part at which the morbid agents were introduced, or the tissues at the seat of inoculation may serve as a focus in which the bacteria germinate and are subsequently disseminated through the system generally, producing profound disturbances and not infrequently death.

Before studying in detail the effects of the introduction of bacteria into an organism, it will be necessary to review the leading facts connected with the evolution of the inflammatory process as manifested by a complex organism.

The simplest animal known—the amœba—is a microscopic mass of nucleated protoplasm capable of spontaneous movement and possessing the power of ingesting, and digesting when suitable, particles of matter presented to it. The essential difference between the simple amœba and the most complex animals is that the latter are compound amœbæ in which individual cells perform separate duties; there is a differentiation of labour; some persist in virtue of their contractility, others for digestive functions; some secrete, others serve for reproduction, and so forth. Most complex organisms are pervaded by a corpusculated fluid, which may circulate throughout the organism by traversing lacunar spaces, or by means of narrow tubular passages possessing distinct walls. This circulating fluid, named blood, serves as a living medium of communication between the various parts of an organism. The blood in higher metazoa contains two kinds of corpuscles; the more numerous are circular or elliptical microscopical discs, tinged of a pale red colour, and, in some vertebrates, furnished with a nucleus. The second variety are nucleated, irregular, colourless, and exhibit amœboid movement; they change their shape, and can escape from the confines of the capillary vessels when such are present. Like an amœba they can ingest, and when suitable digest, particles of matter presented to them. In all vertebrata, with the exception of amphioxus and ascidians, red and colourless corpuscles are present. The invertebrata possess, with very few exceptions, only the colourless corpuscles. These corpuscles, or leucocytes as they are called, fulfil some very extraordinary functions. Should a portion of an animal die, the leucocytes will attack it and, if it be small, will cluster around and by a process of intracellular digestion devour it. When the part to be removed is large, the leucocytes will effect a separation between it and the living body. Not only dead or damaged portions of tissue are thus removed by leucocytes, but useless parts, such as the tails and gills of tadpoles, remains of larval organs, and the tails of ascidians, are thus slowly removed. No animal tissue is capable of resisting an attack of leucocytes. For instance, an examination of the milk-teeth of children or puppies at the time they are shed, well attests the digestive powers of these cells. Surprise is often expressed that when such teeth tumble, or are dislodged, from the gums only the crown is present, the root, or fang, is usually absent. That portion of the tooth in contact with the gum is irregular, and an ordinary magnifying-glass shows it to be full of bays and recesses. When such a tooth is decalcified and suitably prepared for microscopical scrutiny these bays are found filled with leucocytes which, during life, were busily engaged in destroying the fang of the tooth and have slowly induced its fall. The shedding of milk-teeth in mammals, like the disappearance of the tadpole's tail, is due to the persistent attacks of leucocytes. Introduce into the tissues of a cat, a dog, or a man, small pieces of clean sponge. In the course of a few days the fragments of sponge will have disappeared. Vary the experiment by removing the sponge two days after its insertion, cut sections and examine under the microscope, the interstices will be found occupied by an army of leucocytes. Introduce some indigestible object such as glass, a needle, or a fragment of metal. When these are free from any dirty particles and are not lodged in a vital organ, the leucocytes at once attack them, but find such intruders unconquerable; the result is that very large numbers of these cells surround them and gradually become transformed into neutral tissue, thus isolating the intruding bodies from neighbouring parts. The tissue thus formed is known as fibrous-tissue, and the process is termed encystment. Should the intruded body contain particles of dirt offensive to the leucocytes, the action becomes intensified and often disastrous to the cells, for they die in the conflict, and in a few hours the foreign body is surrounded by fluid containing the dead cells. This fluid is usually of a yellowish colour and is known technically as pus; a collection of pus is termed an abscess. As long as an offensive foreign body remains in the organism the abscess enlarges until it finds its way to a free surface and discharges itself: with the evacuation of the pus the cause of the disturbance often escapes.

In their behaviour to foreign bodies leucocytes remind us of the resentment of bees to insects intruding into their hives. When the offender is small it is quickly stung to death and turned out; when large, and they succeed in depriving it of life, it may be too heavy to admit of removal, and the bees render the dead organism inocuous by a covering of wax.

This aggressive behaviour of leucocytes to foreign bodies is extended to such unwelcome guests as pathogenic bacteria. When micro-organisms effect an entrance into an animal the leucocytes attack and attempt to destroy them, and the details of such amœbic warfare may be described from attacks actually witnessed by Metschnikoff in the water-flea *Daphnia*. This observer kept many of these interesting transparent creatures in a tank, and noticed that they became affected with spores which gained an entrance into the body of the crustacean, germinated, and were dispersed by the blood over the body (in *Daphnia* the blood circulates in lacunar spaces) and deposited in those parts where the blood moves slowest, viz., in the cephalic and hinder portions of the mantle cavity: in these places heaps of conidia collect. In the meantime the leucocytes do not remain idle against the invasion, but attack and devour the conidia, take them into their interior and digest them. If a conidium be too much for one cell others join it, form a giant-cell, and thus struggle with the invader. Should the leucocytes overpower the spores, the daphnia lives; if not, the conidia overrun the crustacean and death is the result.

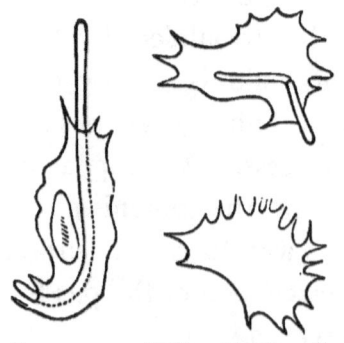

FIG. 114. — White blood-cells (leucocytes) attacking bacilli. (After Metschnikoff.)

A similar process takes place in animals more highly organized, and as no disease illustrates more thoroughly the defending power exercised by leucocytes than that known as avian tuberculosis, the leading points in this widespread affection will be briefly considered. Tuberculosis in man is unfortunately very prevalent, but in birds, especially those which live on grain, it is more common than in human beings. On examining a bird

CAUSES OF DISEASE.

which has died of this disease we find the liver and intestines presenting numerous rounded nodules of a pale yellow colour, varying in size from a pin point to that of a filbert. On cutting into the larger nodules the centre is found occupied by pus. The smaller ones are homogeneous, and on examining them microscopically we recognize in the centre small circular cells with larger ones—giant-cells—lodged among them, outside these a layer of smaller cells with no giant-cells, and lastly a layer of fibrous tissue.

When such specimens are suitably stained, minute rod-like bodies — bacilli — are seen clustered in the centre of the mass and occupying the interior of the cells, especially the giant-cells. In nodules of moderate size the centre is occupied by caseous material surrounded by a zone of cells containing bacilli.

FIG. 115.—Leucocytes ingesting bacilli.

Adjacent nodules may coalesce and thus produce large masses. The blood-vessels connected with the nodules frequently present clusters of bacilli in their interior. An extensive and prolonged study of this disease has convinced me that the bacilli, from whatever source arising, are introduced into the alimentary canal and find their way into the walls of the bowel. Here they are attacked by the leucocytes which surround, ingest, and destroy them. The bacilli may be too numerous for the leucocytes, and the point where the bacilli gain entrance into the tissues become a battle-field, large numbers of leucocytes are quickly on the spot, and reinforce their comrades; as a result of this encounter many of the leucocytes die, others fuse together and form giant-cells: the dead leucocytes form pus cells and give rise to the caseous centre in the nodules, whilst along its confines, in the bacilliferous zone, the conflict continues to rage. The giant-cells are powerful antagonists, for I have seen one contain as many as fifty bacilli. From these nodules the bacilli are conveyed by blood-vessels, or even carried away by the leucocytes, and initiate new struggles in distant parts. It must also be remembered that after their introduction into the body the bacilli will, if the conditions of the host be favourable, multiply very rapidly, and in due course overrun the whole system; nodules arise in the liver, lungs, brain, and skin; function is interfered with and death results. In addition to the local troubles the invasion of an organism by bacteria produces general disturbances, one of the most important being an increase of the temperature of the body, usually termed fever.

It must be borne in mind that local lesions are not necessary results of the entrance of pathogenic organisms into the system. In such a disease as anthrax we have one local sore indicating the seat of inoculation, but beyond the presence of the bacilli in the blood we have no special tissue-change enabling us to identify the nature of the disease, the general disturbance and fever in anthrax conforming to that characteristic of acute, specific, contagious maladies in general.

The behaviour of leucocytes to pathogenic bacteria constitutes the essence of the inflammatory process. It

CAUSES OF DISEASE.

has been known since the time of Celsus that the cardinal signs of inflammation in a warm-blooded animal are redness, swelling, heat, and pain. The redness is due to afflux of blood, the swelling to an increased quantity of fluid in the part, the excess of heat is consequent on the extra tissue-change, and the pain to pressure on the nerves of the inflamed part. The ingenuity of pathologists has devised plans whereby the inflammatory process can be actually watched in such situations as the web of the frog's foot, in the tongue of the frog, and in the mesentery of the mouse. One of the most striking events seen on irritating the parts either by acid, by foreign bodies, or the introduction of bacteria, is the emigration of leucocytes from the walls of the vessels. How the leucocytes escape from the capillaries is a mystery, but that they make their way through the vessel-wall is one of the best ascertained facts of experimental pathology.

The emigrated leucocytes then proceed to attack the intruding matter, and usually effect its removal, but failing this may encyst it in the way already explained. When the invaders are bacilli they may overrun the organism by gaining entrance into the circulatory system. The experimental evidence, and our better knowledge of intra-cellular digestion, shows clearly that, zoologically considered, inflammation is in essence a local struggle between irritants and the white cells of the blood. When the whole of the blood is engaged in the struggle, as in ague, pyæmia, anthrax, and the like, we have general inflammation or fever. The different varieties of fever, when due to micro-organisms, depend on the habits of the specific bacteria; some are more virulent, others are slower in attaining maturity, or are more irritating to the tissues.

This view of the nature of inflammation is of some importance, for it shows that the process may take place in any compound cellular organism, and it certainly takes place in plants; for example, the galls on leaves due to the deposition of eggs in their interstices by insects; each insect producing in this way a different variety of gall, so that one leaf may present at the same time several varieties of galls.

The modification of the inflammatory process according to the nature of the irritant is of interest in its bearing on the evolution of specific diseases. The course of a specific contagious disease, whether due to bacterial infection or to noxious agents produced by bacteria, presents well-marked stages. The first is known as the incubation stage, which may vary from a few hours to days, or even weeks: this is succeeded by a stage of eruption, in which the body presents manifestations of the presence of noxious particles, such as a rash in small-pox, scarlet fever, measles, and the like, or a sore limited to a small area of the body, as in glanders, anthrax, syphilis, &c.

These outward signs are accompanied by fever and general disturbance, marking the maturation of the poison in the organism. After a variable period the signs gradually subside, or terminate the life of the individual, or the bacteria continuing to thrive in the organism may, with the maturation of successive generations, produce periodical disturbance at intervals of several hours, or

CAUSES OF DISEASE. 223

days, as is seen in the various forms of malarial fevers.

The variations in the periods of incubation, eruption, and crisis, correspond with the life-history of the various micro-organisms as ascertained by means of experiments. When bacteria are introduced into cultivation-media, such as solutions of gelatine, agar-agar, chicken-broth, and the like, there is always a latent period which varies with different bacteria. When the micro-organisms commence to grow it is often possible to identify its nature from the mode of growth and behaviour towards the medium, independently of its microscopical characters.

Inasmuch as the varieties of bacteria present intrinsic differences in this respect, it is not unreasonable to suppose that the variations in the period of incubation, character of the eruption, and effects upon an organism should vary according to the nature of the parasite introduced.

Bacteria, like other forms of life, present extreme variety, and the differences between innocent and malignant forms of bacteria, in so far as their effects upon an organism is concerned, are very great. Nevertheless, we can pass by insensible gradations from one form to the other: indeed, the history of micro-organisms shows clearly enough, as in the case of animal parasites, that pathogenic bacteria have been slowly evolved from non-pathogenic forms, and have slowly acquired the power of flourishing upon living bodies when the environment is suitable. This, of course, raises the all-important question, What is suitable environment? Micro-organisms exhibit wonderful vitality and seem to be omnipresent.

Let us take yeast for example. Torula spores exist in countless myriads, and it is necessary merely to expose a solution of sugar to the air in a warm place and in the course of a few hours it is filled with torulæ; according to the amount of sugar present in the solution depends the degree of fermentation established by them, and the amount of alcohol resulting from their activity. The time occupied in producing it depends largely upon the favouring influence of temperature.

Thus it is clear that in this case the determining conditions are the presence of a solution of sugar exposed to air, and a suitable temperature. The spores of the yeast being always around us ready to develop as soon as environing conditions are favourable. So with pathogenic organisms, it by no means follows because they gain entrance into an organism they necessarily flourish. Pasteur demonstrated in the case of silkworms that the micro-organisms found in cases of *flacherie* are only to be found among the pounded leaves in their alimentary canal when the worms digested badly; when the digestive functions of silkworms are active the germs of the micro-organisms are either digested or hindered in their development.

Similar conditions may be studied in man. It is now clear that the disease known as erysipelas is due to pathogenic bacteria which gain entrance into the body through abrasions of the skin. This is more likely to happen when individuals with wounds are surrounded by insanitary conditions, are badly fed, and crowded together. It by no means follows that every wound exposed to the poison of erysipelas necessarily becomes affected, and pathologists are ignorant of the actual conditions of wounds favourable to the development of the micro-organism characteristic of this disease, and it is equally certain that three or more individuals with open wounds may be equally exposed to the virus and yet some of them escape. This immunity may depend on chemical, or fermentative changes, going on in the wound, which produce a medium favourable to the growth and development of the erysipelas-germ. On the other hand it is possible that atmospheric and thermal conditions may favour their development.

The more these questions are studied the more we perceive that the outbreak of infectious diseases depends not so much upon the presence of micro-organisms— for, like the torula, they seem to exist everywhere— as upon the existence of suitable conditions, and as yeast cannot grow and multiply without sugar, neither can the poison of erysipelas, typhus, relapsing fever, and the like, propagate without the presence of some substance produced in living bodies, of the nature of which we are ignorant. This is well shown in Pasteur's researches on fowl cholera: in this instance the microbe would not live in the ordinary cultivation-media employed by him, but when introduced into chicken-broth grew rapidly. On a similar principle relapsing fever is unknown except in times of famine, when the body-chemistry is deranged by want of food, privation, and hardships of every kind.

There is yet another remarkable process which is a modification of inflammation, viz., the repair of wounds. When a wound is made in the tissues of an animal, and it survives the injury, a series of events, varying according to the extent of tissue damaged, ensues. The simplest case is when the tissues are severed with a clean, sharp knife. In this case, unless a large artery is wounded, if the two surfaces of the wound be brought together and maintained in strict apposition, the bleeding will cease, and in the course of a few hours the whole of the damaged surface will be invaded by leucocytes, these gradually elongate and become transformed into tissue and form a uniting medium between the two surfaces of the wound.

When a wound is thus inflicted, and the surfaces not brought into contact, the result is different. Exposure to the air gradually arrests the bleeding, unless large vessels are cut; as in the preceding case the damaged tissue is invaded by leucocytes, and many cells lying exposed on the surface, being too far removed from the living tissue, die. The surface of the wound is at first moist, due to the exudation of fluid from the divided tissues; this, in conjunction with the blood clot, dries in consequence of exposure to the air and forms a scab. This is of great use, for in many instances it hermetically seals the wound, preventing the entrance of micro-organisms. Under the scab the leucocytes unite, become transformed into tissue of repair, and finally, completely fill the gap. Should micro-organisms possessing pathogenic properties gain entrance the leucocytes attack them, the inflammation becomes often intense, fermentation, due to the growth of micro-organisms in the juices of the wound, takes place, products noxious to the individual are produced, and consequences, often disastrous, arise.

In a general work of this character it is impossible to describe in technical detail the different stages and variations displayed in the healing of wounds, but the principles of this important process are the same as those which underlie inflammation. It certainly simplifies our notions of morbid processes to find that the phenomena known as repair of wounds, inflammation, and fever, are manifestations of the same process by which a child loses its milk-teeth, the tadpole its tail, or the stag its antlers, rather than to look upon such conditions as the result of some special law.

CHAPTER XI.

TUMOURS AND CANCERS.

TUMOURS are very interesting to the evolutionist, and in order to obtain a clear notion of them it will be necessary to classify the various "swellings" to which the term is applied, and this is more essential as even medical men use the word in a very indefinite sense. The term "tumour," which literally signifies "a swelling," comprises Cysts, Infective Tumours, Sarcomata, Neoplasms, and Cancers. Each requires separate consideration.

CYSTS.—*A cyst is a tumour containing fluid or semi-fluid contents resulting from the dilatation of a pre-existing cavity.*

Tumours conforming to this definition arise in different ways. For instance, many organs such as the kidneys, liver, salivary glands, and the like, are furnished with a duct, or series of ducts, whereby the fluid secreted by them is conveniently discharged. If from any cause the fluid be prevented from escaping after it has been secreted, it will distend the ducts until they become dilated into large reservoirs, or retention-cysts, as they are termed. This is well illustrated in the case of the kidneys represented in fig. 116. The drawing shows the kidneys with their ducts (ureters) and the urinary bladder of a terrier. The bladder contains two

TUMOURS AND CANCERS. 229

large and two small calculi. One calculus is lodged in the lower end of the left ureter, obstructing it. In consequence of this, the retained fluid has dilated the ureter, and the pressure has induced absorption of the secreting

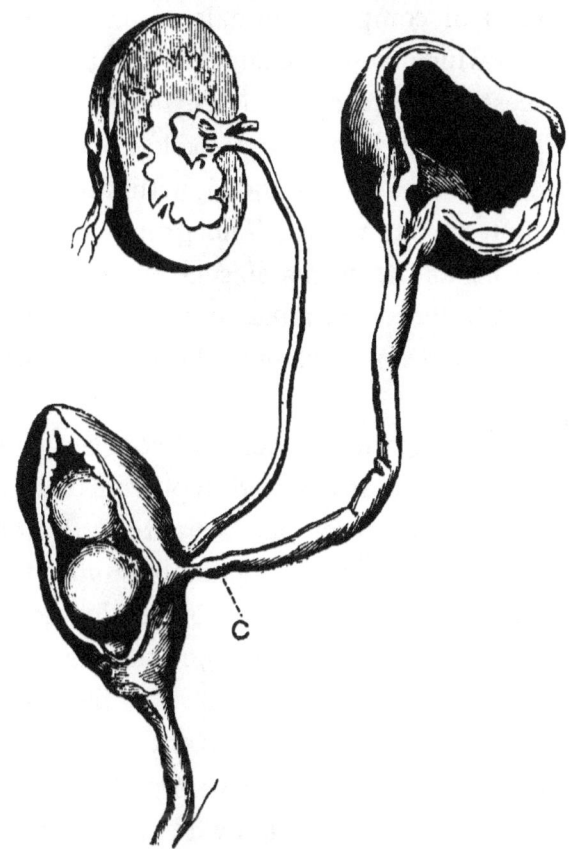

FIG. 116.—The kidneys, ureters, and bladder of a Terrier. The bladder contains calculi, one of which, C, obstructs the left ureter, and has induced a cystic condition of the kidney. (Mus. Royal College of Surgeons.)

tissue of the corresponding kidney, which has become transformed into a hollow bag, or cyst, filled with fluid. This condition of the kidney will arise from any cause obstructing the free flow of urine.

The bodies of animals contain cavities unprovided with ducts : occasionally fluid exudes into such cavities and dilates them into tumours. Such are termed exudation cysts. These have little interest for us.

The bodies of complex animals contain many ducts and passages which were presumably functional in their remote ancestors, but are functionless in existing forms. Such are collectively known as obsolete canals. These canals are occasionally distended with fluid and form tumours often of large size. This group is of great importance to us, as the germs of such tumours are due to modifications induced in animal organization mainly by change in the environment, and inheritance of structural variations.

A very striking example occurs in connection with the ovary. In many mammals and reptiles a collection of vestigial tubular structures has been observed in connection with this organ. In the higher mammals they consist of a series of short vertical tubules dipping into the hilum of the ovary ; above, they end blindly in a duct. This vestigial tubular organ is known as the parovarium, and in the male it forms an important part of the excretory apparatus of the reproductive organs.

From some cause quite unknown to us, one or other of these tubules may become distended into tumours holding many ounces, or even pints, of fluid. Such tumours may jeopardise life from mechanical reasons or induce death by reason of secondary changes in their interior. Cysts of this kind are met with in mammals of all kinds, in birds, reptiles, and in amphibians.

Hen birds occasionally furnish an admirable example

of this kind of cyst. I have already described the vestige of the right oviduct which may be detected on the right side of the cloaca (see p. 64). This slender and apparently innocent tube may now and then dilate to form a cyst as large as a walnut (fig. 117). As a rule such cysts are harmless, but at times they inflame and become filled with pus; should the cyst rupture the pus escapes among the intestines and the bird dies from peritonitis.

Other cystic conditions arising in functionless ducts are described in Chapter III.

In addition to cystic tumours arising from retention of fluid, or in functionless ducts, there is an interesting class known as diverticula, or false cysts: these demand some consideration.

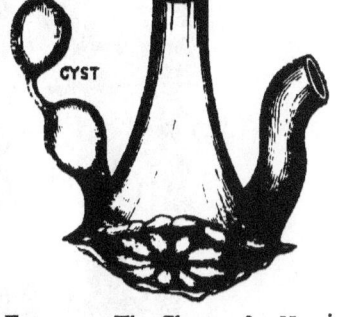

FIG. 117.—The Cloaca of a Hen; the stump of the right oviduct is dilated into a cyst.

Many of the tubes and passages of animal bodies possess two walls, an outer one, more or less rigid, which gives it shape, and an inner one lined with a soft velvety covering known as mucous membrane. This differs from the outer tube in being usually distensible and easily separable from the outer or protecting wall. Not infrequently from strain, accumulations of fluid or air, the inner, moveable, elastic mucous membrane lining such a tube will be forced through a weak or defective spot in the outer tube and form a soft rounded bulging or diverticulum; the cavity of such a diverticulum still retains its connection with the tube to which it belongs.

232   EVOLUTION AND DISEASE.

One of the most remarkable diverticula known to me exists in the neck of the Emu (*Dromæus novæ-hollandiæ*), which demands consideration, as it serves as an admirable physiological type of diverticula in general.

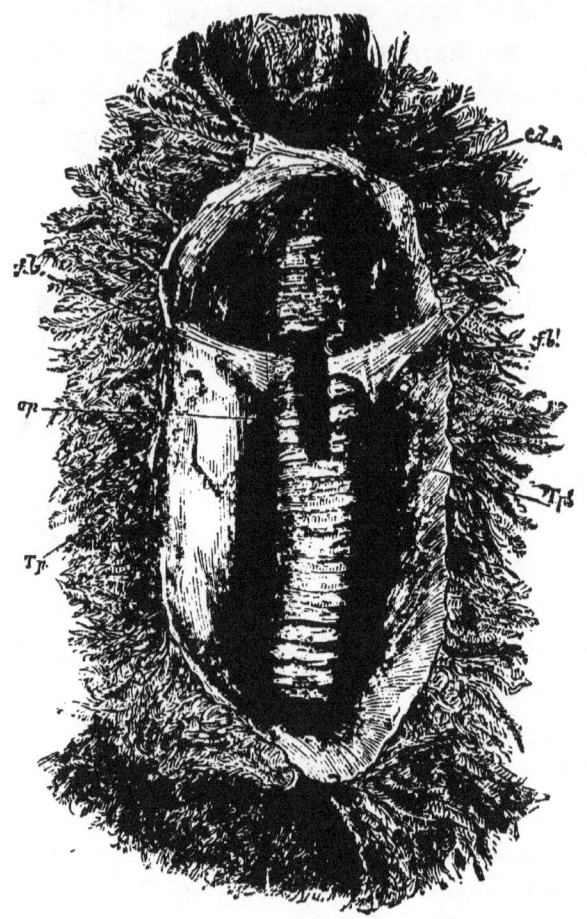

FIG. 118.—Tracheal pouch of the Emu.   (After Murie.)

In the emu there is a natural defect in the trachea at a spot varying in position between the fiftieth and sixty-fifth rings. This spot has the form of an oblong slit, which in the adult bird is about seven centimetres in length; the width varies from eight to twelve millimetres, according to the retracted or distended condition of the windpipe. The mucous membrane lining the trachea passes through this slit, and forms a large bag or sac, which lies between the trachea and the skin (fig. 118). During the breeding season, the birds inflate the sac with air and produce a peculiar drumming sound which resembles the booming noise made by blowing forcibly across the opening of a large but narrow-necked bottle This cyst is not present in the emu chick, only the slit in the trachea; as the bird grows, so the mucous membrane becomes slowly protruded through it. On one occasion I have found this cyst filled with fluid mucus, due to inflammation of its walls, and on attempting to relieve it, the mucus passed into the trachea and literally drowned the bird.

Among diverticula, analogous to the tracheal pouch of the emu, should be mentioned the singular pouch of the Bustard (*Otis tarda*).

INFECTIVE TUMOURS.—Tumours belonging to this group are caused by micro-organisms. They demand close attention, because they are the most generalized of all tumours and occur in every kind of vertebrate animal.

Structurally, they consist of small round, or spindle-shaped cells, intermixed with giant-cells in variable proportions. Infective tumours are of two classes: (*a*) sarcomata, (*b*) infective granulomata. A sarcoma usually appears as a tumour, and later infects the system, producing secondary nodules in different organs, such as the lungs, liver, &c. The infective granulomata appear as small scattered nodules in various parts of the body. In many cases the micro-organism which produces the disease, has been satisfactorily isolated. A few tumours have been placed in this group because their structure, history, and infective properties correspond to those in which micro-organisms have been satisfactorily detected. The consideration of infective tumours belongs strictly to the evolutionist, for the causative agents may be regarded as parasites which have gradually acquired the power of flourishing in animal bodies. This group may be illustrated by an account of the remarkable disease known as actinomycosis.

Actinomycosis is a disease prevalent among cattle. It commences as a rule in connection with the jaws and tongue in the form of nodules, which become confluent and form large masses. After attaining a certain size these tumours suppurate and discharge pus. When the jaw is the seat of the disease, the bone becomes eroded and expanded on account of the growth invading it. Such tumours were formerly described by veterinary surgeons as sarcomata, but in 1877 Bollinger detected in them microscopic organisms of peculiar radiating structure, termed in consequence actinomyces, or the rayed-fungus. The matter has received the close attention its interest and importance deserves. It appears that the spores of the fungus enter the tissues, either through ulcers, decayed teeth, or the sockets of teeth, and acting as irritants establish inflammation. After lodging in the tissues for a time, the spores develop a mycelium (fig. 119), and the radiated structures thus formed become surrounded by leucocytes, round cells, and giant-cells,

TUMOURS AND CANCERS. 235

which form perceptible nodules. Many nodules becoming confluent give rise to a tumour in the affected part. That the actinomyces is the cause of the disease is demonstrated by the fact that the fungus may be cultivated artificially, and when introduced into a calf experimentally will produce the disease.

Actinomycosis is interesting pathologically, but is also important from an economical point of view, and still

FIG. 119.—A tuft of Actinomyces highly magnified, showing the clubs.

more important in a sanitary respect, as quite a number of cases have been placed on record in this country, but more especially in Germany, which have occurred in the human subject, and it is a noteworthy fact that in many of the patients the disease seems to have commenced in the alimentary canal.

Sarcomata.—Those tumours which pathologists term sarcomata differ from those produced by the ray-fungus in the following particulars. The micro-organism or causative agent has not yet been isolated, and we have no satisfactory evidence that a sarcoma can be inoculated into another animal. Nevertheless the two forms of tumours agree in the general principle of structure, disastrous effects upon the life of the individual, and in a tendency to infect the system. Careful research will probably establish before very long a poison or micro-organism for each of the various types of sarcoma. My own inquiries into these tumours has long served to convince me that such will be the case, and a few of the reasons will be briefly detailed. Every day experience teaches that tumours in the human subject are extremely common. Attendance at a veterinary infirmary will soon convince a regular visitor that tumours are frequent in horses, cattle, sheep, and dogs. A long and careful personal attendance at many thousand *post-mortem* examinations of wild animals, dying in captivity, has disclosed the fact that such animals are rarely affected with tumours. A critical analysis of facts further shows that in man cancer is more common than infective tumours. In domesticated mammals cancer, in the sense in which it will be employed later, is unusual, whilst infective tumours are extremely common. In wild animals nearly all the tumours belong to the infective granulomata, only a few cases of cancer being known. It may be useful to detail one or two typical specimens of sarcomata from animals.

The first is a round-celled sarcoma growing in the subcutaneous tissues of the neck of a hen. It is of the size of a chestnut, and is surrounded by a capsule of fibrous tissue. On section it has a pale yellow colour, and is elastic to the touch. When portions of the tumour were hardened, and thin sections prepared for the microscope, it was found to be made up of a multitude of closely-packed round cells, with here and there slender fibrillæ of delicate tissue passing between them; occasionally a giant-cell, with many nuclei, was seen. The

FIG. 120.—The head of a Fowl, with a sarcoma growing in the subcutaneous tissue.

general appearance of the tumour may be inferred from the drawing in fig. 120. In this case the sarcoma grew in the subcutaneous tissue, and was of small size; but such tumours may grow in any situation of the body, sometimes in bones, where they attain a very large size; in the brain, eye, intestine, limbs, &c. I have examined tumours in fish, frogs, birds, snakes, marsupials, rodents, carnivora, quadrumana, and ruminants.

When sarcomata grow from bone, especially from the interior of a bone, they usually possess large numbers of giant-cells. When originating in pigmented spots such as the black or pigment coat of the eye, or the pigment layer of the skin, they are of a deep black colour, and named in consequence melanotic. Grey horses are especially liable to this form of tumour, yet we have no reason to believe that the coloured races of mankind are more or even so prone to them as Europeans.

Sarcomata do not always remain localized in this way. After the tumour has been growing for a time, other nodules make their appearance in different parts of the body, and not infrequently the secondary formations are larger than the original tumour. These facts, and the general effects of such tumours, would alone cause us to suspect some parasitic agent, and what is of utmost importance, the early removal of the primary tumour occasionally prevents general infection.

In the chapter on Inflammation, the relation of the leucocytes to bacterial invasion was described. Let us ascertain how it will elucidate the nature of sarcomata. The classification of these tumours is founded on the structural characters displayed by thin sections of the dead tumours under the microscope; they are then described as round-cell tumours, spindle-celled, melanotic or giant-celled. The appearance of a section of a round-celled sarcoma is exhibited in fig. 121.

When fluid portions of such tumours are examined,

whilst they are yet alive, on a suitable stage, these cells have been found to exhibit amœboid movements and change of shape. When such cells die they assume a rounded form, in the same way that a dead leucocyte becomes transformed from an irregular shapeless mass of active protoplasm to a definite rounded cell.

To put the matter in a clear form, a sarcoma is probably the scene of action of a violent and prolonged

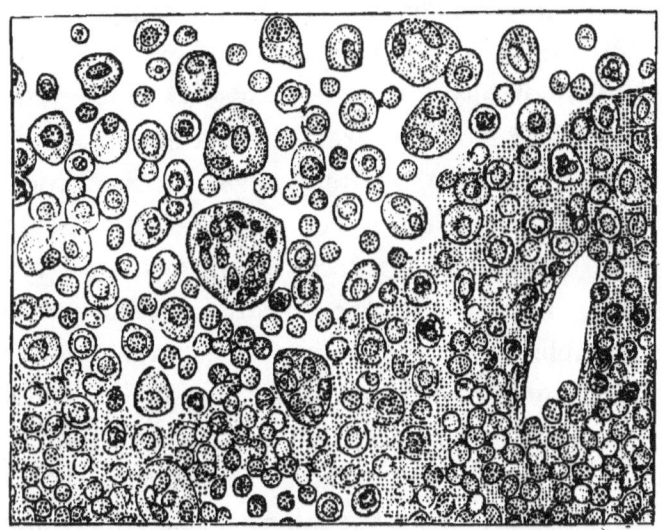

FIG. 121.—The microscopic appearance of a round-celled Sarcoma.

conflict between irritant micro-organisms and leucocytes. I say probably, because, as has been already remarked, bacteriologists have not yet succeeded in isolating a special bacterium for sarcomata in general; that such agents will soon be discovered is in the highest degree probable, because in recent years each increase in the list of infective granulomata is made at the expense of sarcomata. The structure, mode of growth, infective properties, and manner in which these tumours destroy life, clearly coincide with what is positively known with regard to infective granulomata. The fact that sarcomata make up the greater part of tumours occurring in wild and domesticated animals has, in my opinion, a very significant import in this relation.

NEOPLASMS.—Tumours belonging to this group will not detain us long. They are innocent in so far as the life of the individual is concerned, and are composed of fat, bone, or cartilage; in some cases they consist of an aggregation of blood or lymph vessels. Such tumours may cause inconvenience from their large size, or interfere with vital organs, but they never produce constitutional disturbance or infect the system. In many instances they occur as local overgrowths of tissues, resembling in this way the deviations which occur in the vegetable world and known as "sports;" this term being used by gardeners as signifying a bud or off-shoot which suddenly assumes a new, and sometimes very different, character from that of the rest of the plant. The term "spontaneous variation" is sometimes applied to such conditions. As "sports" occur throughout the plant world, so simple neoplasms occur throughout the vertebrate kingdom, and wherever fat, bone, and cartilage are found, will the tendency to "sports" exist and produce fatty tumours, bony tumours, cartilaginous tumours, and the like.

Cohnheim attempted to account for the occurrence of neoplasms by supposing that during the development of an animal a certain number of the original cells of a part remained undeveloped, and that later in life they grew erratically or aimlessly and formed tumours. This view has been discussed critically by almost all writers on tumours since Cohnheim expressed it, and it undoubtedly accounts for many neoplasms. The great objection to the view has been that such undeveloped rudiments have not been shown to exist. More careful researches show clearly enough that among the great class of morbid productions generically referred to as tumours, Cohnheim's theory holds good for cysts, many neoplasms, and a remarkable group known as dermoids, and by a careful extension of the definition "tumour germ," it could be applied to cancer. Restricting the application of this theory to the tumours indicated, this view offers adequate explanation of bony and cartilaginous neoplasms, of some vascular tumours, and those which have been already referred to as arising in connection with vestigial structures. (See Chapters III. and IV.)

The full details cannot here be discussed, but any one exercising patience in such anatomical inquiries will soon be able to satisfy himself, as I have done, that "tumour-germs" actually exist in our bodies, and of such a character as Cohnheim's theory requires. The erratic growth of such undeveloped portions of tissue may be well illustrated in a simple way by examples from the vegetable kingdom. The stems of trees and woody plants form a large number of buds, most of which grow into branches. Some of these remain undeveloped for a time, and then, instead of forming a normal branch, they grow erratically, and form a swelling or woody tumour of irregular shape, which may attain a large size. Such a tumour of a tree is termed a xyloma. The bud-like character of such woody tumours is shown in an interesting series presented to the museum of the Royal College of Surgeons by Mr. Stephen Paget. From some of the tumours buds have formed, and in one case the bud has grown into a minute branch. Every swelling on a tree, however, is not a woody tumour or xyloma; many are due to the irritation of insects.

CANCERS.—We have now to consider the tumours whose main structural peculiarity is that they contain epithelium. The group is of great importance in that it includes the terrible disease known as cancer. It is only of late years that the term cancer has come to possess any strictly scientific significance. In the early days of pathological anatomy any tumour presenting malignant characters was termed cancer, but in the present day the term is restricted to tumours structurally resembling imperfectly formed glands. In order to appreciate the nature of cancer it will be advantageous for us to briefly study the evolution of glands in general. I can only attempt to give in abstract the large amount of evidence I have accumulated, in order to show that cancers are aberrant glandular formations, and may not inaptly be defined as "biological weeds."

In complex animals the free surface of the body and the alimentary canal is covered with cells differing from those found in the underlying tissues. Such cells are known collectively as epithelium, and though varying in shape in different situations and under various conditions, present identifying characters. This epithelium is prone to dip below the surface and invade the underlying tissues. In their simplest form such downgrowths are at first solid and club-shaped. Subsequently the central cells liquefy whilst the peripheral ones arrange themselves in a definite and regular order, so as to form a lining membrane to the central chamber, or acinus. The portion of the acinus near the surface from whence the downgrowth originated is slightly narrowed and constitutes the duct of the gland. This is one of the simplest

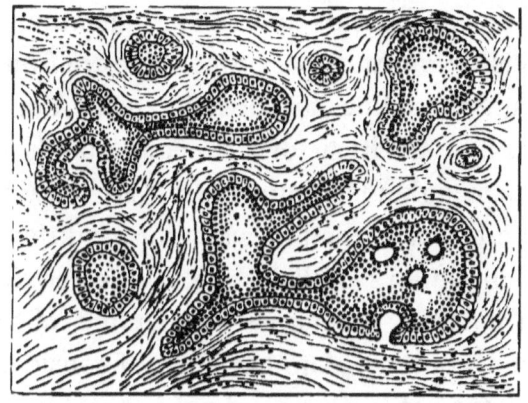

FIG. 122.—The microscopical appearance of an epithelial Tumour (Cancer).

forms of gland, and from it any complex type may be derived by buddings or outgrowths from the primary epithelial germ (fig. 122). Secreting glands are arranged by anatomists into groups according to their structure. Physiologists give them specific names according to the secretion furnished, such as mucus, sweat, milk, and the like. Structurally they form two main groups, tubular and racemose. The simple tubular gland is met with in the intestine of man and in any mammals, whilst the racemose variety is represented by sebaceous glands usually attached to hairs, but may occur in situations where hairs are not found normally. A number of tubular glands may be collected together and give rise to a compound organ like the kidney; in the same way a collection of racemose glands may form a compound organ, such as the salivary glands, or the poison glands of snakes. In complex animals like vertebrata, glands of peculiar character are restricted to definite parts of the body. Thus sebaceous glands are almost peculiar to the skin ; mucous glands to mucous membrane, and so forth. Thus it comes to pass that there are frontier lines, and as a rule these lines are not violated. For instance, at the lips the territory of sebaceous glands ends and that of mucous glands commences. At the termination of the large bowel or rectum, a similar condition of things exists, Lieberkühn's follicles so characteristic of the bowels terminate. Numerous frontier lines of this nature exist in our bodies. Another fact of considerable importance is that at birth an animal is not furnished with all the glands it will require, the germs of many arise in after-life and even subsequent to the years of active growth. In the chapter on supernumerary mammæ, it was pointed out that in recesses of skin or mucous membrane glands grow luxuriously, and to this may be added, especially if the recess is moist.

In young individuals we find occasionally in connection with a functional gland, a tumour which when examined microscopically displays all the features peculiar to the gland with which it was connected ; the only point in which it differs is that the adventitious mass is impotent, that is, it cannot produce the secretion peculiar to the gland from which it arose. Such a tumour is called an adenoma, and receives a specific name according to the gland it resembles—sebaceous, mammary, renal, hepatic, &c. Adenomata may attain enormous size and weigh many pounds. As life advances the mimicry is crude, the cells, instead of clothing the alveoli in a regular manner, are tumbled together in confusion. Such tumours are cancers; they grow aimlessly, having no function to keep them in subjection, and being poorly supplied with blood vessels, undergo degenerative changes, and the cells being dispersed over the body may reproduce, in remote tissues and organs, secondary tumours resembling the original cancer from which they arose.

The glandular nature of cancers is further illustrated by the fact that in their intimate structure they resemble the glands in the immediate neighbourhood. Thus a cancer of the lip resembles the cutaneous glands; in the liver it mimics the liver; mammary cancer resembles imperfectly the secreting tissue of the breast, and so forth.

Many competent pathologists are of opinion that cancers like the infective tumours are due to a microorganism; this is very probable, although thus far inquiries in this direction have not yet succeeded in identifying such agents; nor is cancer inoculable from one animal to another. Should a bacterium be ultimately found as the causative agent, it will in no way affect the arrangement of cancers into a group apart from other tumours, as they exhibit in such a marked degree the glandular type of structure which alone serves to distinguish them from sarcomata, with which they were frequently confounded, nor is the distinction always made with the accuracy that is essential in order to enable us to draw conclusions from statistical tables as to the frequency of cancer in comparison with other tumours.

Another circumstance which has never received full attention is the great frequency of cancer in the white races of mankind. A careful and extensive inquiry into this question is very desirable, for we are grossly ignorant as to the occurrence of cancer among the natives of colonies, even of India.

Again, cancer, using the term in the sense in which it is employed throughout this chapter, is rare among domesticated mammals, and rarer still among wild mammals, even those living in captivity. To take one example, cancer of the uterus, which is responsible for the death of an appalling number of women every year in England, is, as far as my inquiries have extended among veterinarians, as well as from my own observations, very rare in domesticated and wild mammals.

The cause of the extreme frequency of cancer in one case, and its absence in another, may be in a measure explained by certain peculiarities in structure and gland distribution. The question is far too extensive to be adequately considered in such a work as this, and when I have pushed my investigations further I hope to deal with it in a systematic manner.

Before leaving this subject it will be interesting to describe a specimen illustrating the close relation which exists between glands and cancer. In the section devoted to supernumerary mammæ, I drew attention to the relation existing between cutaneous recesses and glan-

TUMOURS AND CANCERS. 247

dular organs, and used this as an explanation of the inordinate number of teats found within the pouch of some opossums. It is a remarkable fact that one of the most typical specimens of cancer that has come under my notice in a wild animal occurred in a short-headed

FIG. 123.—The posterior half of a short-headed Phalanger. The pouch is occupied by a cancer. (Nat. size.)

phalanger (*Belideus breviceps*) (fig. 123). In this case the pouch was occupied by a tumour as big as the kernel of a filbert, and when we remember that the parts of the phalanger are represented of natural size in the drawing, the tumour was relatively large. In order to appreciate the significance of this specimen, it should be studied in conjunction with the remarks on the marsupial pouch contained on p. 171, *et seq*.

SUMMARY.—To epitomise the facts briefly considered in this chapter relative to tumours, from an evolutionist's point of view, it may be stated that (excluding those arising from accidental interference with the function of secreting organs), some tumours arise as a result of change of function in organs, rendering some part of them useless; others arise from the introduction into an organism of bacteria which have by imperceptible stages become slowly modified so as to be able to thrive on or in animal bodies; some, and these the most innocent forms, arise as "sports"; whilst the gland tumours and cancers are due to the epithelial modifications which give rise to secreting glands. It may be further stated that animals other than man are liable to tumours, agreeing in all respects with those which have been so long and closely studied in him. Of all tumours occurring in the lower animals, so far as the facts at our disposal show, the commonest forms are the infective granulomata, including sarcomata.

With respect to the cancer group there are some very extraordinary facts which cannot be passed over in silence. For instance, hair, teeth, feathers, glands of all kinds and cancers arise on fundamentally the same plan,—a downgrowth of epithelium into the subjacent tissues; it is certainly a suggestive fact that abnormal irritation will produce a crop of hairs in an unusual situation, as was shown in an early chapter, and used to explain the curious hairs at the pyloric end of the darter's stomach. In the same way glands may be induced to grow by irritation, moisture, and warmth combined, and a continuance of such conditions will often provoke an outbreak of cancer.

CHAPTER XII.

THE ZOOLOGICAL DISTRIBUTION OF DISEASE.

A LITTLE reflection soon convinces us that as the habits, structure, and environment of animals differ very widely, the manifestations of disease must also vary. Little, however, has been done in the direction of studying the zoological distribution of disease, and its consideration in this work may be regarded as premature. Still it may be useful to indicate the amount of information we possess on the subject, and its scanty proportions should serve as a stimulus for further inquiry in this direction, and show how necessary it is that those who have had opportunities of making observations on this subject should record their experience. The matter is rendered more difficult from the impossibility of obtaining positive information concerning the diseases of wild animals in a state of nature; even the difficulty of obtaining their bodies is illustrated, in the case of monkeys, by Dr. Falconer in explaining the paucity of the remains of quadrumana in geological strata. "When the monkey pays the debt of nature his carcass falls to the ground, and immediately becomes the prey of the numerous predaceous scavengers of torrid regions, the hyæna and wolf. So speedily does this occur, that in India, where monkeys occupy large societies in mango groves around villages, unmolested and cherished by man, the traces of casualties among them are so rarely seen, that the simple Hindoo believes that they bury their dead by night."

As far as my own investigations have extended I find, excluding the affections known collectively as the acute exanthemata (scarlet-fever, measles, small-pox, and the like), that most diseases known in the human species occur in mammals. A few affections rare in man are frequent in mammals; a limited number of diseases are peculiar to mankind, whilst others occur only in the lower animals. The inquiry is of some importance, for it serves to show that certain diseases give rise to changes so very different in animals of one class to those of another class as to be described under a different name, whilst two distinct affections may produce in animals belonging to different classes lesions of such close naked-eye resemblance that they are frequently confounded with each other. Such conditions raise the all-important question, "How far is it probable that many of the acute contagious fevers which affect the human species occur in other animals, but producing different symptoms receive another name?" This is illustrated in a remarkable manner by tuberculosis, a disease of world-wide distribution. Writing concerning that common manifestation of this disease, pulmonary consumption, Hirsch, in his admirable *Geographical and Historical Pathology*, states that it has held at all times and among all civilized peoples a foremost place among the national diseases, and that it extends over every part of the habitable globe, and may be designated ubiquitous in the strictest meaning of the term.

In man tubercle affects any organ and tissue of the body, and is due to the entrance and multiplication in the body of a micro-organism, the tubercle bacillus, identified by the genius of Koch. Until the discovery of this bacillus in tubercular lesions it was customary to apply the term tubercle to almost any disease which was characterized by the formation of nodules in the tissues, thus the term came to have a generic rather than a specific signification. Now that we have a pathological criterion of tubercle, the term tuberculosis has a definite meaning, and certain affections formerly included are now known to be due to other causes, and numerous affections formerly excluded now help to swell the list of troubles due to this omnipresent micro-organism. Among other mammals the disease has a peculiar distribution; it is very common among cattle under the name of grape disease, or its German equivalent, *Perlsucht*. Monkeys living in confinement in this country are occasionally attacked by it, but not so frequently as was formerly supposed. Among grain-eating birds the disease is a perfect scourge; the flesh-eating birds are not so liable to contract it, and are probably not infrequently attacked in consequence of devouring tubercular grain-eating birds. In quadrumana and man the disease runs a similar course, whilst in cattle the lesions are so different, that it would be difficult to believe that it is in any way related to the tuberculosis of *Primates*, were it not for the existence of identical microorganisms, and this again applies equally to birds in whom the lesions differ from those in man and cattle. It is also extremely difficult to understand the immunity of horses, tuberculosis being rare among *Equidæ*.

Anthrax demonstrates in a remarkable manner why disease should have a zoological distribution, for slight physiological differences protect an animal against the action of the anthrax bacillus, the morbific agent of the ruinous splenic fever. This disease can easily be communicated to the ox, sheep, rabbit, and guinea pig by injecting into the circulation a small quantity of blood taken from an animal which has died of splenic fever. Such injections are rapidly fatal. On the other hand, it is difficult to inoculate the dog and pig, and fowls never acquire the disease. The cause of the immunity of fowls has been cleverly explained by Pasteur. It had been ascertained that the anthrax bacillus does not develop when subjected to a temperature of 44° Centigrade. The body temperature of a fowl is about 41° C., whilst that of the horse is 37·7 C., the dog and rabbit, 38°–39° C. On immersing the feet of a fowl in cold water at a temperature of 25° Cent., so as to reduce its body heat to 37° or 38°, and then injecting it with blood from a case of splenic fever, it was found at the end of twenty-four hours dead, with its blood filled with the bacteria of splenic fever. In another experiment a hen was inoculated and subjected to the cold-water treatment; when the fever was at its height the hen was taken out of the water, wrapped carefully in cotton wool and placed in an oven at 35° C. In the course of a few hours it was restored to health. Hens killed after being experimented upon in this way exhibit no trace of the bacteria in their blood.

Under ordinary conditions a frog cannot be killed by injection of anthrax cultures, but if, after inoculating a frog, its temperature be raised to 36° by carefully warming the water in which the frog is placed it will succumb.

Such facts as these throw great light upon the restriction of diseases to particular groups of animals, and it explains the readiness with which the tubercle bacillus flourishes in man, for experimentally it has been found to develop most luxuriously at a temperature of 37° to 39° Cent. This would serve to explain the rarity of tubercular lesions in cold-blooded animals.

Up to the present time tubercle has only twice been recorded in reptiles. The first specimen I observed in a large Python (*Python molurus*). The nodules in the various organs contained bacilli in large numbers. In this instance I am of opinion that the reptile contracted the disease from eating tubercular birds. Mr. W. K. Sibley reported a case of tuberculosis which he found in a snake (*Tropidonotus natrix*). As the tubercle bacillus flourishes at a temperature of 37°–39° Cent. it at first seems difficult to account for tubercular lesions in snakes. In a valuable series of observations made by Mr. Forbes[1] on an incubating python at the Zoological Gardens, the temperature of the male was found to vary from 28°–30° C.; the temperature of the female under the same conditions of external warmth was 29°–31·6° C. These observations were made in July, and the greatest temperature recorded between the folds of the male was 32° C.; for the female, 33·8° C. It is also of interest to find that the temperature of the pythons, taken between the folds, was higher than the surrounding air, sometimes as much as 6·4° C. in the

[1] "Collected Papers, 1885," p. 285.

male, and 9·2° C. in the female. Valenciennes, in some similar observations conducted in 1841 in the Jardin des Plantes, Paris, found the temperature of the incubating python as high as 41·5° C.

The observations are of interest, for it indicates the occasional possibility of the python's body temperature rising sufficiently high to favour the development of the tubercle bacillus, and as the python's temperature appears to be slightly raised above that of the surrounding media, it would come very close to the required 37° C. when the snake was exposed to the full glare of a hot midsummer sun.

Under such conditions a snake when exposed to tubercular food resembles a European when exposed to the dangers of malaria on an unhealthy tropical coast. Vagary, in the liability to or immunity from a special disease among closely allied families of mammals, is exhibited in other than infectious diseases. Take, for instance, gout. No one has ever clearly shown that this affection occurs in animals other than man. It is stated that parrots are liable to gout, but this question assumes a different aspect when studied in relation with an interesting disease of the hog known as guanin gout.

In man, apart from the pain and disturbance induced by an attack of gout, we find deposited in the less vascular parts, such as cartilage, tendinous and fibrous tissues, masses of a crystalline nitrogenous substance known as urate of soda. The crystals are needle-shaped, and in severe cases form collections, familiar to those who have very gouty relations, as chalk-stones. These deposits constitute the most constant pathological condition in gout.

There is a parasitic affection common in swine (and not infrequently found in man), to which attention has of late years been largely directed, known as trichinosis. The parasite *Trichina spiralis* when hatched finds its way into the voluntary muscles of man and the pig, there becomes encysted, and in due course is surrounded by calcareous particles. The encysted worms are visible to the naked eye in cut sections of muscle as small dots. In 1866 Virchow detected in a piece of ham some small white concretions which were regarded as trichinæ; but on examination were found to be of crystalline structure, and to furnish the reaction for guanin, a crystalline nitrogenous body resulting from chemical changes in animal tissues, first discovered by Unger in Peruvian guano.

Guanin seems to be very widely distributed in the animal kingdom; it occurs in fish, the excrement of spiders, in the pond mussel, the pancreas and liver of the horse, and the skins of frogs and lizards. It responds to easily applied chemical tests. The interest of guanin for us centres itself in the fact that it produces lesions identical in their pathological anatomy with gouty lesions, that is to say, it becomes deposited in cartilage and fibrous tissue, forming deposits exactly resembling the urate of soda deposits in man. Such resemblance is not confined to naked-eye characters, but extends also to the microscopic details as is shown in the drawing (fig. 124), taken from Dr. Mendelson's admirable contribution to this interesting subject. When examined in

ZOOLOGICAL DISTRIBUTION OF DISEASE. 257

thin sections under the microscope, the affected tissues are found impregnated with feathery tufts of crystals, which respond to the ordinary tests for guanin.

From the hog we may turn to parrots, which frequently present in their subcutaneous tissues, cartilages, skin, muscles, and intestines, nodules, which, in their naked-eye characters, are indistinguishable from gouty nodules

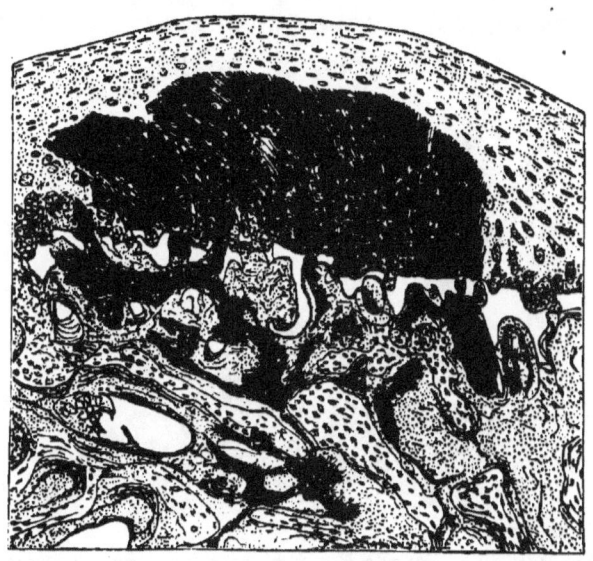

FIG. 124.—Microscopic section of the articular cartilage of a pig's knee-joint affected with guanin gout. (After Mendelson.)

as seen in man. Indeed such deposits have been described as gout, not only in parrots, but in fowls, falcons, ostriches, &c. In the drawing (fig. 125), the foot of a parrot is represented with characteristic nodules, and when cut into, the centre of such nodules is occupied by a white mortary-looking substance in every way resembling that known as chalk (urate of soda) in human gout. The analogy goes beyond mere naked-eye characters, for these deposits consist of a crystalline substance resembling guanin and urate of soda, and responding to the murexide test. Guanin and urate of soda are closely allied bodies, and as far as I can ascertain it is no easy task to decide between them. Until the subject has been more thoroughly investigated it will remain uncertain whether the nodular lesions of parrots and other birds are really of the same nature as gout in man, or allied to guanin gout of the hog, and it is very

FIG. 125.—The feet of a Parrot with supposed gouty nodules.

probable that some case of supposed uratic gout in man may be guanin gout.

This confounding of avian and human gout, and trichinosis with the guanin gout of the hog, is easy to comprehend, as it requires chemical cunning to complete the identity, but we must remember that true gout was formerly, and is now, often confounded with a disease of the articular ends of bone, which, though often referred to as rheumatic gout, has no relation with gout, except that both diseases are distinguished by nodules on the joint ends of bones. Rheumatic gout, or, as it is now more appropriately termed, osteo-arthritis, is a disease of great interest, for it has a wide zoological distribution, and has even been detected in the joints of the extinct Irish Elk (*Megaceros hibernicus*). This disease is characterized by enlargement of the ends of bone, destruction of the cartilage and synovial membrane of joints, with calcification of the ligaments. It is no respecter of persons; young and old, rich and poor, high and low, suffer from it. I have detected the disease in the joints of a snake's backbone, in birds of various kinds, including the neck of an ostrich, in cats, dogs, leopards, lions, tigers, horses especially, oxen, sheep, kangaroos, bears, and many others. It is as far as I can ascertain the most widely diffused of all the bone diseases to which vertebrated animals are liable.

In order to show the care necessary in such generalizations we may take the recent additions to our knowledge of such a long recognized disease as cretinism. All visitors to Switzerland, the Rhone and Aosta valleys, are familiar with what is termed endemic goître and endemic cretinism. The leading features of cretinism are briefly these:—The disease is congenital, and displays itself in unnatural shortness of the trunk and limbs, malformations of the skull (as a rule it is unusually small), and idiocy, combined with abnormal conditions of the thyroid body. Cretins, as those affected with this disease are called, present a characteristic appearance; a typical cretin is represented in fig. 126, taken from the admirable report, compiled by a Commission, created by the King of Sardinia, to inquire into this disease, published at Turin, 1848. The boy is described as being twenty years old, is one metre in height; he is deaf and dumb, not comprehending any sign, and spends his time in turning the small bâton between his fingers; the intellectual faculties are practically wanting.

Fig. 126.—An endemic Cretin twenty years old: one metre high. (From the Sardinian Report.)

Until recently it was little thought that cretinism occurs in our country, although it was known that goître is endemic in certain districts. In 1871 Dr. Hilton Fagge[1] clearly showed that cretins occurred in England

[1] "Medico-Chir. Trans.," 1871.

ZOOLOGICAL DISTRIBUTION OF DISEASE. 261

by publishing descriptions of undoubted examples. One of Fagge's specimens is drawn in fig. 127, for comparison with the endemic cretin on the opposite page. This boy is sixteen years and a half old, rather less than a metre in height, and could understand a good deal of what was said to him, ask for what he wanted, and the parents could understand what he said.

FIG. 127.—A sporadic English Cretin, rather less than a metre in height, and sixteen years of age. (After Hilton Fagge.)

In subsequent investigations Hilton Fagge showed that, contrary to what he had originally stated, the thyroid body in some of these children was abnormally large.

Since the attention of medical men has been drawn

262 EVOLUTION AND DISEASE.

to this matter, a large number of cretins has been detected in England, and many have been critically observed, and detailed accounts of their anatomy placed on record, which remove any doubts as to the identity of the English sporadic with the Alpine endemic cretin. It is also significant that since careful accounts of the leading features of the disease have been circulated, cretins have been recognized in many parts of England, and instead of being limited to the small village of Chiselborough, in Somersetshire, it turns out to be a far from infrequent condition in many large towns,

FIG. 128.—A Calf-cretin. Length of trunk, thirty centimetres. Length of limbs, five centimetres.

including London. Cretinism is not confined to the human species. In 1877 H. Müller described a cretinous calf, and subsequently Eberth was able, in a monograph on this subject, to refer to cases which had been reported in the human subject under different names. Thus far cretinism has been recorded several times in the calf, in sheep, and dogs, and the careful accounts of the anatomy of the specimens leave no room for doubt as to its identity with the cretinism of man. Animal cretins occur not only in regions where the disease is endemic in man, but also in England,

and a typical specimen of a calf-cretin which came under my observation is sketched in fig. 128. This calf exhibits in a striking manner the leading features of the affection. For instance, its trunk only measures thirty centimetres in length, and the legs are five centimetres long: the head arrests attention on account of its shortness, resembling strongly the head of a pug-dog. Cretinism is not unknown among dogs; the museum of the Royal College of Surgeons possesses an excellent specimen in a fœtal puppy.

The calf-cretin is of interest especially with regard to the undue shortness of the head and limbs, for it has been suggested that the pug-dogs, which are such favourite pets with many ladies, are cretins, and that by selected breeding a race of cretinous dogs has been produced. It has also been suggested that the short-legged spitze or Dachshund is possibly cretinous. This however is problematical; the spitze must be a very old variety of dog. Dr. Blackmore showed me in the Salisbury museum some arm bones which clearly belonged to a dog, and they were curved, or bowed like the bones of a spitze; the curves were certainly not due to rickety changes. These bones were obtained among others from excavations made in investigating the pit-dwellings, admirable models of which are exhibited ir the same museum.

Dr. Parrot has made a careful study of cretinism, and allied diseases of the skeletons, and goes so far as to believe that the ancient Egyptians were acquainted with cretinism, and even had a cretinous god, Ptah, which was particularly venerated at Memphis. An examination of the models and figures of this god preserved in the Egyptian galleries of the British Museum shows, that in some of the figures Ptah is represented as a big-bellied, squat divinity, with short

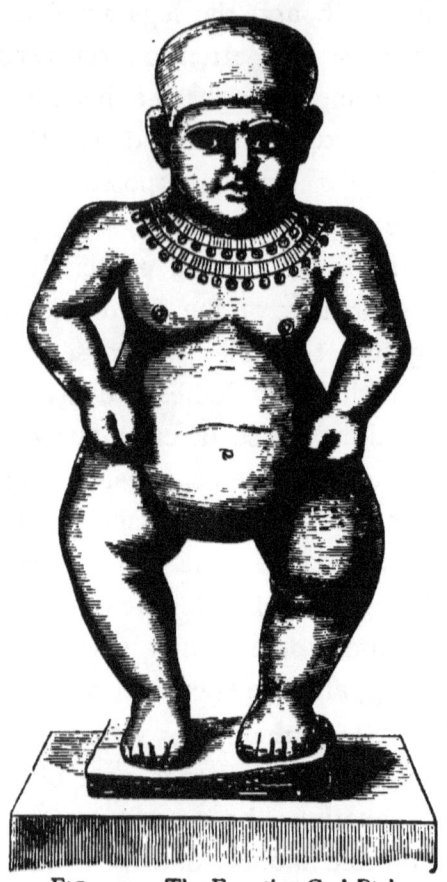

FIG. 129.—The Egyptian God Ptah.

limbs, not unlike a cretin (fig. 129), in others he is represented as of noble figure, the type of lordly bearing and majesty. The notion that the dwarf models of Ptah represent cretins is very speculative and improbable.

Cretinism has a wider distribution, geographically and zoologically, than we were aware even twenty years ago, and it is not unreasonable to suppose that careful inquiry will show that many other similar diseases, supposed to be rare or confined to certain districts, are as a matter of fact commoner than we suspect.

Rickets is another example of a disease having a wide zoological distribution. The leading characters of rickets are, undue softness of bone in young animals, associated with catarrh of the stomach and intestines, depending upon, or induced by, unsuitable food and unfavourable surroundings. The softness of the different parts of the skeleton gives rise to a complicated series of deformities, some of which are incompatible with life.

For a long period rickets was supposed to be restricted in its distribution to England, and is, to this day, often referred to as the English disease. Now, we know that rickets occurs all over Europe and in other parts of the world, and recently I had an opportunity of examining portions of a rickety skull which were obtained from Lamoo, an island near Zanzibar; it was found buried in sand, on the site of an old battle-field, by Dr. Briscoe.

Rickets is so very common in man that *a priori* we should expect it to be frequent in other members of man's class. On this head we possess but scanty information. Mammalian skeletons have been preserved in museums, and erroneously labelled; we now know that most of them are rickety skeletons, for the systematic inquiry which I have conducted into the diseases of mammals dying in the Zoological Society's Gardens has established the fact that rickets is a very common disease. Indeed, so frequent is it among quadrumana that half the monkeys and lemurs brought to this country die rickety. This disease affects, besides men, chimpanzees, orangs, gibbons, macaques, baboons, capuchins, squirrel and spider monkeys, and lemurs. Among carnivora we find it in lions, tigers, hyænas, bears of all kinds, and in the domestic cat, the dog, fox, raccoon, and seal. Among ruminants it occurs in deer, sheep, and goats. In rodents, the beaver, porcupine, rabbit, and coypu rat are affected by it. Among marsupials, the kangaroos, phalangers, and opossums are most liable. In birds it has been found in the emu, ostrich, rhea, and pigeon.

All who have studied the disease are of opinion that it is due to deficiency of lime salts with the food. Mr. T. D. A. Cockerell has argued, and I think on good grounds, that the scalariform shells of some mollusks may be regarded as arising from the same cause as rickets in vertebrates. Thus rickets has an exceedingly wide zoological distribution.

Those singular productions known as cutaneous horns are interesting in connection with the subject matter of this chapter. In the Introduction some remarks were made concerning such horns in relation to physiological types, but the question was by no means exhausted.

Man, in common with many mammals, possesses glands in the skin, which secrete an unctuous material known as sebum. Such glands are termed sebaceous, and are more abundant in certain regions of the skin, whilst in other parts they are absent. Not infrequently changes occur in the interior of such glands causing the sebum to be retained, and the acini of the glands

FIG. 130.—A Mouse with a cutaneous horn arising in a sebaceous cyst.

becoming dilated by the accumulated secretion forms a perceptible swelling known as a sebaceous cyst. At times the secretion bursts through the cyst-wall in the form of a dark-coloured corneous substance, which increasing in length, by additions at the base, may attain a length of many centimetres. Such a horn is shown in fig. 130, growing from a cyst on the back of a mouse, and in order to show its relation to the sebaceous cyst, the horn is shown in vertical section in fig. 131.

FIG. 131.—The horn and cyst in section.

Cutaneous horns, in their naked eye and even microscopical characters agreeing with those formed in sebaceous cysts, may arise from the transformation of warts, and so closely do the two forms mimic each other that it is not easy to distinguish between them. Thus to all outward appearance the horn of the mouse in fig. 132 is identical in its structure and naked-eye characters to that growing from the back of the mouse ; but I

Fig. 132.—The head of a Mouse with a cutaneous horn arising from a wart. (Mus. Royal College of Surgeons.)

have had ample opportunity of demonstrating its origin in a wart. These specimens illustrate how necessary it is to exercise care in discriminating between pathological productions apparently resembling one another.

Cutaneous horns are very common in man, and have been known to attain a length of fifteen centimetres ; they were formerly regarded by ignorant persons as

ZOOLOGICAL DISTRIBUTION OF DISEASE. 269

objects for veneration. In addition to the modes of origin mentioned, horns may arise from the overgrowth of nails, and also in the scars of burns; these have no connection with our present study.

Horns arising in sebaceous cysts, or from warts, have been recorded in man, oxen, sheep (fig. 133), goats, dogs,

FIG. 133.—The head of a Sheep with a large wart-horn on its head.

and hare. I have recently seen one growing near the hock of an eland: in birds they have been recorded in the parrot, greenfinch, linnet, canary, and oyster-catcher. The museum of the Royal College of Surgeons, London, contains an admirable collection of such horns, one of them which grew from the flank of a ram, and from its structure is clearly a wart-horn, is the largest on record, it measures in length nearly one metre, and is twenty-eight centimetres in circumference at the base.

In birds cutaneous horns grow very rapidly, and are shed with each moult: in the course of three months such horns have been known to attain a length of seven

Fig. 134.—The head and leg of a Thrush: the horns on the leg arise from warts, that on the head from a sebaceous cyst.

and even ten centimetres, and to be reproduced for several consecutive years. An excellent specimen of this form is shown on the thrush (fig. 134). This bird was observed by Mr. Roger Williams: it had three wart horns on the legs, and upon its head a sebaceous cyst with a cutaneous horn just commencing to protrude from it. The bird is interesting as illustrating the two modes by which these horns arise. The statement that birds have no sebaceous glands except the uropygial gland requires modification.

Cutaneous horns, due to the hardening of secretion, occur as normal productions in several animals. Thus the patch of spines on the fore-arm of hapalemur (*H. griseus*) and on the fore-arm of the ring-tailed lemur (*L. catta*), are of this character. Even more curious are the wart-like processes formed on the skin of the thigh in lizards, from the hardening of the secretion furnished by the femoral glands; this hardened secretion enables the male to clasp the female.

Among the few diseases restricted to mankind must be mentioned leprosy. *Elephantiasis græcorum*, as true leprosy is called, is a very remarkable affection : it has never been seen in any animal other than man, although determined attempts have been made to communicate it to rabbits, monkeys, cats, dogs, and fowls. Neisser claims to have successfully produced leprous tubercles in rabbits and dogs, but the results have not been confirmed by others who have repeated his experiments. The latest contribution to this subject is by Dr. Beaven Rake, Superintendent of the Trinidad Leper Asylum. The possibility of causing leprosy by inoculation has occupied his attention during four years. He has performed fifty-four experiments, some being the direct introduction of the diseased tissues from man into the subcutaneous tissues of guinea-pigs and rabbits, and by feeding fowls on leprous material. In all cases the experiments failed to produce constitutional leprosy.

Leprosy is distinguished anatomically by the formation of nodules or tubercles in the skin, mucous membranes and underlying tissues. When the skin is affected the hands, feet, and face are most frequently attacked. The nodules commence as red spots in the skin, which become gradually of a blue tint, then brown: the subjacent tissue becomes thick and hard. The tubercle

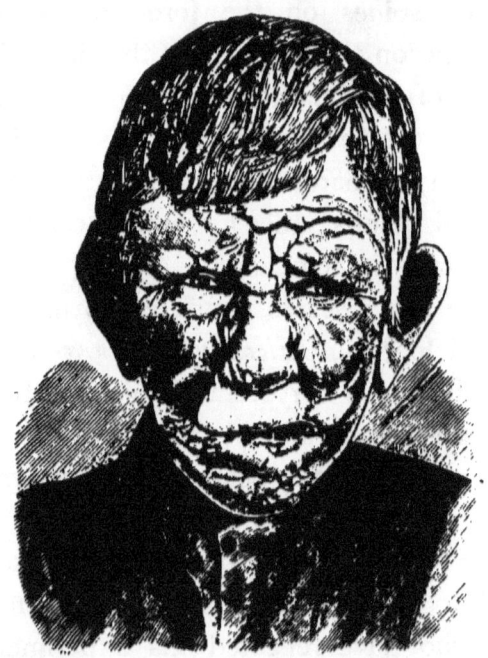

FIG. 135.—The face of a Leper. (After Ziegler.)

increases and forms a sub-globular, soft, pale-coloured prominence, which, when knocked or rubbed, may ulcerate. Leprous ulcers are commonest on the inside of the nose, on the conjunctiva, and mucous membrane of the larynx. When the nodules form on nerve sheaths anæsthesia is produced.

Microscopically, leprosy nodules are found to be made

up of round cells, with larger ones known as leprous cells intermixed. When suitably stained the nodules are found to contain enormous numbers of bacilli, many of which crowd the large leprous cells.

It should be borne in mind that in the Mosaic books the term leprosy is used in a generic sense, for it included many forms of curable skin disease under this name, as a careful perusal of that very interesting chapter (Leviticus xiii.) will clearly show. It was only by isolating and carefully watching the initial red spots that a confident opinion could be expressed. The evidence furnished by the same chapter indicates that even when the disease was well pronounced, other affections than that which we now recognize as *Elephantiasis græcorum* were classed together under the term leprosy, but leprosy in animals is not mentioned once in the whole Pentateuch.

The means described in the Mosaic books of isolating suspected cases, and allowing time to settle the diagnosis, are admirable. Even in these days of advanced civilization well-trained and thoughtful physicians find that the isolation of infectious cases is the best preventative treatment known. It is more than probable that this old-established custom explains the present restriction of leprosy to definite regions, such as Norway, Finland, and the Baltic provinces of Russia, Central and Southern America, South Africa, and Asia.

As far as our knowledge of the zoological distribution of disease at present extends there are two affections peculiar to mankind, viz., true leprosy and syphilis. Even mental affections occur in animals : we have already seen that the form of idiocy known as cretinism is recognized in calves, dogs, and lambs. Horses are certainly known to suffer attacks of acute mania, and dogs become demented and are occasionally imbecile, not recognizing their master from other persons, and forgetting their own names. From being keen at sport, they fail to recognize a hare or partridge, even blunder over them in a field and take no notice of the game when it rises. Convulsions caused by troubles during dentition are well recognized in dogs, as well as a nerve affection, closely similar to epilepsy. The remarkable affection of human beings termed locomotor ataxy, and the singular St. Vitus' dance, or chorea, occur in dogs.

It will, perhaps, be desirable to conclude this chapter by a brief consideration of the strange disease known as *Elephantiasis arabum*, the zoological and geographical distribution of which is limited in an interesting manner by mosquitoes.

In India, China, Egypt, Arabia, the West Indies, and Australia this disease is endemic : it is characterized by enlargement of the parts affected, which may be the arms, legs, and other parts of the body. The skin covering the part becomes enormously thickened, forms hard masses and folds obscuring the toes when the legs are affected, producing an appearance not unlike an elephant's leg. One of the most remarkable features in this extraordinary disease is that very frequently the blood of individuals with elephantiasis swarms with minute hæmatozoa, named by Lewis *Filaria sanguinis hominis;* and recent researches go to show that the presence of filaria in the blood in association with

ZOOLOGICAL DISTRIBUTION OF DISEASE. 275

elephantiasis is so frequent that they are manifestations of the same disease.

According to the admirable researches of Dr. Manson the chief facts connected with this disease are the following :—

The adult worm is probably taken into the alimentary canal of man in drinking-water. From the stomach it bores its way into the thoracic duct or some lymphatic vessel, and is subsequently joined by one of the opposite sex. Here they may live for years, discharging their embryos into the lymph stream, to become distributed by the blood current over the body.

One of the most remarkable phenomena connected with filarial disease is that the embryos disappear from, and reappear in, the general blood stream at certain periods of the day. Under ordinary conditions the blood of an affected individual presents no filariæ between the hours of nine a.m. and six p.m. ; after six p.m. they begin to make their appearance and increase in numbers until midnight, at which hour as many as 260 filaria have been counted in a single drop of blood : from this hour they gradually diminish in numbers and at nine o'clock in the morning they cannot be detected in the blood. During the night, and whilst the filarial migration is at its height, mosquitoes visit the body, and Manson has identified the female, a particular variety of *Culex* which, by the structure of its oral appendage, abstracts blood from the filariated individual, be it man or beast. As the culex selects eight o'clock in the evening as the feeding hour it necessarily follows that the embryo filariæ are taken into the stomach of the mosquito, and Manson by a series of well-planned experiments, has demonstrated that this insect acts as the intermediate host of the parasite. During the following four or six days the embryos complete their metamorphosis, and the friendly mosquito has completed its life cycle and dies, and its body probably falls into the water in which the eggs were deposited. The filaria by this time has developed so far that it is capable of living independently of the mosquito, and it seems probable that it remains in the water until it is captured by some animal in search of food, or swallowed by man, thus enabling it to complete its development.

These facts indicate conclusively that this disease is spread and, in all probability, maintained by mosquitoes, and the geographical distribution of the disease is coincident with that of culex. In this connection it is a fact of some importance to remember that filariæ occur in other animals as well as in man. Manson states that half the dogs in China, all the magpies, one-third of the crows, and many other birds harbour similar hæmatozoa in prodigious numbers; and in South China where the filarial disease is endemic, if the blood of one thousand natives, selected indiscriminately, be examined some time between sunset and sunrise, in about one hundred the *Filaria sanguinis hominis* will be discovered.

The point of interest to us in connection with this disease is that though the filariæ occur in dogs, magpies, and I have detected them in the blood of macaque monkeys, the enormous enlargement of the legs and other parts of the body, which is one of the chief characters of the disease in man, has not been recorded in these animals. This, again, is another example of the same cause producing varying effects in different species of animals, of which we have had several instances in this chapter, and I can only repeat that if this subject were carefully and broadly investigated, many diseases supposed to be distinct in man and the lower animals would be found to depend upon the same cause.

INDICES.

INDEX TO AUTHORS CITED.

Ahlfeld, 144
Albers, 144
Albrecht, 71, 111
Alston, 155
Altum, 155
Aristotle, 31

Baker, M., 132
Balfour, 41
Bardeleben, 161
Barth, 170
Bartlett, A. D., 205
Bell, 195
Bidder, 22
Blackmore, 263
Blumenbach, 206
Briscoe, 265
Bruce, M., 167

Champneys, 171
Cheselden, 202
Cicero, 31
Clarke, 118
Cleland, 130
Cockerell, 266
Cohnheim, 240, 241
Collins, 141

Cuvier, 196, 206

Darwin, 24, 32, 45, 61, 98, 139, 148, 150, 167
Deen, Van, 126
Devey, 193
Descartes, 43
Dugés, 117
Dumas, Blanche, 130

Eberth, 262
Egremont, Lord, 155
Eschricht, 209
Eve, 72

Fagge, Hilton, 260
Falconer, 250
Flower, 10
Forbes, 11, 254
Förster, 144
Freeman, H. W., 190

Garrod, 26
Gaskell, 51
Gegenbaur, 99, 135
Goethe, 139

Godsir, 206, 209
Graaf, 42
Günther, 29
Gurlt, 127

Haeckel, 115
Haswell, 110
Hensel, 159
Hensinger, 84
Hewitt, 150
Hilaire, 25, 120, 144
Hirsch, 251
His, 81, 180, 183, 192
Howes, 89, 139
Humphry, 15
Hunter, 17, 21, 111, 153
Hyrtl, 177

Kleinenberg, 117
Koch, 252
Kowalevski, 49

Lamprey, 197
Lewis, 274
Luschka, 96

Macalister, 196
Malkmus, 172
Manson, 275
Matthews, 46
Mendelson, 257
Milne-Edwards, 139
Müller, 262
Murray, Jardine, 110

Nasymth, 206

Ogle, 136
Oppian, 31
O'Reilly, 197
Owen, 65, 206

Paget, Stephen, 242
Paley, 171
Parker, 100
Parrot, 263
Pasteur, 224, 225, 253
Pliny, 31
Pitt, Percival, 97
Potton, 193

Rake, Beavan, 271
Rauber, 121
Recklinghausen, 57
Rengger, 36

Salvin, 185
Schmidt, 182
Sclater, 166
Sedillot, 98
Shattock, 90, 100
Sibley, W. K., 254
Spencer, 42
Sterling, 22
Stewart, 90

Tiedemann, 26, 103
Tomes, Chas. S., 208, 209, 212
Traquair, 186
Treves, 68
Tuckerman, 128
Turner, 209

Unger, 256

Valenciennes, 255
Virchow, 23, 54, 56, 256

Wallace, 33, 61
Werfer, 197
Windle, 193
Wolff, 143
Woodward, 30
Wyman, 33

INDEX TO ANIMALS.

Alces machlis, 5, 106, 156
Amblyopsis, 33
Amblystoma, 119
Amœba, 214
Amphioxus, 49
Antilocarpa americana, 11
Ascidians, 39, 49
Astarte, 31
Asterias equestris, 103
Ateles paniscus, 15, 159
Aye-aye, 202

Babirussa, 204
Baboon, 19
Belideus breviceps, 247
Beluga leucas, 89
Bos indicus, 170
Bufo vulgaris, 111
Bunting, 157
Bustard, 157, 233

Cæcilia compressicauda, 42
Callithrix, 159
Camel, 19
Capreolus caprœa, 156
Cariacus virginianum, 157
Cavicornia, 10
Ceratodus, 42
Cercopithecus patas, 169
Cervulus, 146
Cervus canadensis, 69
Cervus elephas, 157
Cetaceans, 60, 89
Chætodon, 195
Chaffinch, 157
Cheiromys, 164
Chimpanzee, 99, 136
Chirogaleus coquereli, 163
Cotinga, 157

Crystallodes rigidum, 115
Cuckoo, 157
Culex, 275

Dace, 4
Darter, 26
Didelphys cancrivora, 172
Didelphys virginianum, 172
Dove, 114
Dorking-fowl, 112
Drake, 130
Dromæus novæ-hollandiæ, 104, 232

Echidna, 137, 163
Elasmobranch fish, 76
Elephant, 204
Emu, 104, 232
Equidæ, 16, 252

Filaria sanguinis hominis, 276
Fierasfer, 29
Fratercula corniculata, 11

Gibbon, 66, 99
Goat, 20, 84
Gorilla, 19, 99, 145
Grampus, 209
Ground parrot, 35

Hapalemur griseus, 164
Hæmatopus ostralegus, 10
Hare, 189
Hatteria, 43
Himalayan monaul, 104
Hipparion, 159
Hircus, 178
Hircus thebaicus, 85
Horse, 16

INDICES.

Hylobates leuciscus, 66, 109

Ichthyosaurii, 158

Labyrinthodonta, 43
Lacerta viridis, 124
Lamellibranch, 28
Lemurs, 136
Lemur catta, 163, 271
Lemur macaco, 163, 165
Lemur mongoz, 163
Lepidosteus, 37, 42
Lizard, 124
Lobster, rock, 139
Lophophorus impeyanus, 104
Lumbricus terrestris, 116

Macaque, 99
Macacus sinicus, 168
Macropus, 193
Margarita margaritifera, 29
Marmoset, 159
Megaceros hibernicus, 69, 259
Meleagrina, 31
Menobranchus, 97
Menopoma, 97
Mesoplodon layardi, 209
Metazoa, 215
Microcebus smithi, 163
Mole, 32
Monkeys, American, 136
Moose, 5, 107
Mot-mots, 185
Mustelus, 42
Mycetes, 159

Narwhal, 60

Octopus, 4
Opossum, 172
Orang, 99, 136, 188

Orca, 209
Ornithorhynchus, 60, 137, 145, 163
Ostrich, 19
Otaria gillespii, 89
Otaridæ, 89
Otis tarda, 233
Oyster, 31
Oyster-catcher, 10

Palinurus, 139
Parrot, 59, 258
Patas monkey, 169
Peacock, 157
Pea-crab, 31
Pea-hen, 149
Pearl oyster, 29
Pectunculus, 31
Pelican, 157
Phalanger, 144
Pheasants, 157
Phocidæ, 89
Pithecus satanus, 145
Pinnotheres, 30
Plotus anhinga, 26
Polish fowl, 23
Primates, 252
Prongbuck, 11
Pterodactyl, 144
Puffin, 11
Python molurus, 211, 254

Rana esculenta, 126
Rana palustris, 128
Rana temporaria, 112
Rangifer, 156
Rhinoceros, 9
Roedeer, 7

Seabright bantam, 150
Sea gull, 17

Seal, 89, 145
Sepia, 4
Shark, 76, 121
Sheep, 27
Siamang gibbon, 145
Silkworm, 224
Skate, 46
Sloth, 57
Sphenodon, 43
Spider monkey, 159
Squirrel, flying, 144
Star-fish, 103
Stringops habroptilus, 35

Syphonophora, 116

Tarsius spectrum, 164
Trichina spiralis, 256
Tropidonotus natrix, 254
Trout, tailless, 186

Varanus, 42
Viper, 62

Wapiti deer, 69
Wombat, 202

Zebu, 170

INDEX TO SUBJECTS.

Actinomycosis, 235
Ægipans, 56, 90, 177
Aftershaft, 104
Allantois, 40
Anthrax, 253
Antitragus, 177
Antlers, 5, 107, 155
Astragalus, 100
Atavism, 134
  „ Spurious, 143
Auricle, 63, 79, 83, 177
Auricular fistula, 178
Autosite, 122

Bacteria, 213
  „ pathogenic, 213
Big-toe, 14
Bifid tail, 124
Bone, 5
Bone of attachment, 211
Branchia, 80
Branchial fistula, 81
Bullets in tusks, 208

Cæcum, gibbon, 66

Cæcum, horse, 67
  „ man, 65
  „ tiger, 66
  „ rhinoceros, 68
Calamus, 104
Callosites, 19
Cancer, 242
Cauda helicis, 177
Circumcision, 184
Cleft-palate, 189
Commensalism, 29
Conjunctiva, hair on, 141
Corns, 19
Cretinism, 259
Cretins, 194
Cysts, 228

Defects, acquired, 176
  „ transmitted, 176
Devil, 57
Diana of Ephesus, 162
Dichotomy, artificial, 115
  „ limbs, 102
  „ tail, 124
  „ trunk, 119

Dichotomy in worms, 116
Disuse, 14, 35
Ducts, functionless, 231

Elephantiasis arabum, 274
,,   græcorum, 271

Fauns, 56, 91
Feathers, 105
Fever, 213
Fermentation, 224
Fibula, 97
Fistula, auricular, 177
,,   branchial, 81
Flacherie, 224
Fracture, Pott's, 96
Functionless ducts, 231

Giant-cells, 238
Gill-slits, 80
Gizzard, 17
Gods of the woods, 56
Gout, Guanin, 255
,,   uratic, 255
,,   in parrots, 257
Guttural pouches, 96

Hare-lip, 189
Hearing, 32
Horned men, 197
Horns, 9, 266
Hypertrophy, 17
Hyporachis, 104

Impaction theory, 132
Infective granulomata, 233
Inflammation, 213
Inherited defects, 188
Irritation, 21

Lalloo, 123

Lanugo, 137
Leaves, metamorphosis of, 139
Legs, supernumerary, 127
Leprosy, 271
Leucocytes, 215
Levator costæ muscle, 135
Lobule of pinna, 177

Malformations, 188
Malleoli, 99
Mamma, 162
,,   abdominal, 168
,,   axillary, 171
,,   brachial, 165
,,   inguinal, 163
,,   supernumerary, 167
,,   thoracic, 168
Meningocele, 23
Micrococci, 214
Micro-organisms, 217
Milk-glands, 162
Muscle agitator caudæ, 53
,,   depressor ,, 53
,,   levator, 52
,,   levator costæ, 135

Odontome, 73
Œsophageal collar, 48
Opercula, 64
Organisms, pathogenic, 213
Osteo-dentine, 208
Ovaries, 149
Overgrowth, 20
,,   of beaks, 58
,,   claws, 57
,,   hoofs, 20
,,   nails, 57
,,   spurs, 22
,,   teeth, 199
,,   from disuse, 57
Oviducts, 45, 64

Parasite, 122
Patagium, 144
Patella, 17
Pigment, 3
Pineal body, 50
„ eye, 42
Pinna, 176
„ of whale, 89
Polydactyly in horses, 159
„ in man, 159
„ monkeys, 100
Pre-hallux, 161
Pre-molars of horse, 70
Pre-pollex, 161
Ptah, 263

Repair, 225
Reliquia, 62
Reversion, 134
Rickets, 265
Right aortic arch, 137
Ritta-Christina, 120
Rudimentary organs, 60

Santos, dos, 130
Sarcomata, 235
Satyrs, 56, 90, 177
Sexdigitism, 194
Sexual characters, 148
Solanum jasminoides, 32
Shackle joint, 196
Spina bifida, 23, 56, 75
Spinal cord, 48
Sports, 240
Spurious atavism, 143
Spurs, transplanted, 153
Stomach, hour-glass, 149
Supernumerary antlers, 107
. „ arms, 115
„ auricles, 83
., digits, 108

Supernumerary hands, 110
„ heads, 122
„ legs, 111
„ mammæ, 167
„ nipples, 167
„ rays, 103
„ ribs, 135
„ tails, 124
„ teeth, 106
„ tragus, 179

Tails, true, 52
„ false, 54
Tailless trout, 186
Talipes, 188
Teeth, overgrown, 199
Teratomata, 125
Tongue, 77
Tonsil, 93
Torula, 224
Tracheal pouch, 232
Tragus, 177
Tumours, 228
Tuberculosis, 251
Two-headed Nightingale, 121

Urachus, 41
Use and disuse, 14
Useless parts, 61

Variations, 45
Viscera, Transposition of, 131
Vermiform appendix, 65
Vestigial parts, 60

Warts, 8
Webbed-fingers, 145
„ legs, 143
Wolffian bodies, 45

Yolk-sac, 75

www.ingramcontent.com/pod-product-compliance
Lightning Source LLC
Chambersburg PA
CBHW021958220426
43663CB00007B/868